Answers to your top safety management questions

Andy Tilleard
CMIOSH MIIRSM EurOSHM

OAK·TREE·PRESS

Published by
OAK TREE PRESS
19 Rutland Street, Cork, Ireland
www.oaktreepress.com

© 2010 Andy Tilleard

A catalogue record of this book is
available from the British Library.

ISBN 978 1 904887 41-6

INTRODUCTION

QUICK WIN SAFETY MANAGEMENT is aimed at entrepreneurs and business managers who want to understand how to protect their business investment, meet their legal obligations and cut their costs.

By using a simple, effective and internationally-recognised safety management model, entrepreneurs and business managers can learn the basics of effective safety management, including the principal safety management components, how they fit together and relate to each other and how improvements in safety performance can be made and measured.

QUICK WIN SAFETY MANAGEMENT is designed so that you can dip in and out seeking answers to your top safety management questions, as they arise. Reflecting the International Labour Organization document, *Guidelines on Occupational Safety and Health Management Systems - ILO-OSH 2001,* there are five sections to the book:

- Policy.
- Organisation.
- Planning and implementation.
- Evaluation.
- Action for improvement.

Policy sets out the safety management aims and objectives for your organisation, including how it will meet its legal, financial and moral obligations.

Organisation defines what organisational structures will be put into place in order for your organisation's management and employees to meet the aims and objectives set out in the policy section, including defining management levels, job position responsibilities and accountabilities and reporting hierarchies.

Planning and implementation outlines how the safety management system will be planned and implemented to meet your organisation's policy objectives, including setting measurable objectives and targets, hazard prevention and risk management.

Evaluation outlines the processes in place to evaluate the performance of your organisation's safety management system, to verify whether the aims and objectives set out in the policy section are being met.

Action for improvement shows how your organisation can learn from its own failings or mistakes within its safety management system and, where these have been identified, how improvements can be made.

In addition, using the grid in the **Contents**, you can search for questions and answers across a range of topics, including:

- Communication and training.
- Contractors.
- Definitions.
- Employees.
- Management.
- Systems and procedures.
- Standards.

And, where appropriate, answers cross-reference to other questions for a fuller explanation or more information.

Enjoy the book – I wish you lots of quick wins and success in managing safety in your organisation!

Andy Tilleard
Cork
August 2010

CONTENTS

Search by theme:

Or search by topic:

Communication and training

Contractors

Definitions

Employees

Management

Systems and procedures

Standards

using the grid overleaf.

POLICY

	Communication & training	Contractors	Definitions	Employees	Management	Systems & procedures	Standards	Page
Q1 What is health and safety?			☑					2
Q2 What is a safety management system?			☑		☑	☑		3
Q3 How is a safety management system organised?					☑	☑		5
Q4 What is the background to the *ILO-OSH 2001* safety management system?						☑	☑	8
Q5 What are some of the obstacles to safety management?						☑		10
Q6 What are the most common hazards in the workplace?						☑		12
Q7 What is a policy statement?	☑		☑					13
Q8 What policy statements do we need?	☑				☑	☑		15
Q9 What is the process for writing policy statements?	☑				☑	☑		17
Q10 Who should write our safety management policy statements?	☑				☑	☑		19
Q11 What does an occupational safety and health policy look like?	☑					☑		20
Q12 How do we communicate our occupational safety and health policy statement?	☑					☑		22
Q13 What should be included in additional policy statements?	☑					☑		24

ORGANISATION	Communication & training	Contractors	Definitions	Employees	Management	Systems & procedures	Standards	Page
Q23 How do we format corporate safety documents?	☑				☑	☑		50
Q24 How do we format a work procedure?	☑				☑	☑		52
Q25 What are management procedures?			☑		☑	☑		54
Q26 How do we document management procedures?	☑				☑	☑		55
Q27 What is communication?	☑		☑			☑		59
Q28 What information needs to be communicated within a safety management system and how?	☑				☑	☑		61
Q29 Who should receive health and safety information?	☑	☑		☑	☑	☑		65

PLANNING & IMPLEMENTATION	Communication & training	Contractors	Definitions	Employees	Management	Systems & procedures	Standards	Page
Q30 What is an initial review?			☑		☑	☑		68
Q31 What should be in an initial review?					☑	☑		70
Q32 What is meant by planning, development and implementation?					☑	☑		72

PLANNING & IMPLEMENTATION	Communication & training	Contractors	Definitions	Employees	Management	Systems & procedures	Standards	Page
Q33 What safety standards should we adopt?					☑	☑	☑	73
Q34 What are safety management objectives?			☑		☑	☑		75
Q35 How do we decide what occupational safety and health objectives are required?					☑	☑		77
Q36 When should we review our occupational safety and health objectives?					☑	☑		79
Q37 What is planned preventative maintenance?			☑		☑	☑		81
Q38 What are the key elements of a planned preventative maintenance system?					☑	☑		83
Q39 What can be covered under a planned preventative maintenance system?					☑	☑		85
Q40 What are safety plans?					☑	☑		87
Q41 What is risk assessment?					☑	☑		89
Q42 Why is the risk assessment process so important?					☑	☑		92
Q43 What are the potential problems with risk assessment?					☑	☑		95
Q44 What does 'reasonably practicable' mean?			☑		☑	☑		97

PLANNING & IMPLEMENTATION	Communication & training	Contractors	Definitions	Employees	Management	Systems & procedures	Standards	Page
Q45 What is a hierarchy of control?			☑		☑	☑		99
Q46 How should we use personal protective equipment?					☑	☑		101
Q47 Why is personal protective equipment always at the bottom of a hierarchy of control?					☑	☑		103
Q48 How can we apply hierarchy of control procedures in practice?					☑	☑		106
Q49 How can we manage occupational noise exposure?					☑	☑		109
Q50 How can we manage working-at-height?					☑	☑		111
Q51 What is a hazard register?			☑		☑	☑		113
Q52 What is a permit-to-work system?			☑		☑	☑		116
Q53 What activities require permit-to-work controls?					☑	☑		118
Q54 What is the lock-out / tag-out system and how is it used in permit-to-work activities?			☑		☑	☑		120
Q55 How should we review our permit-to-work system?					☑	☑		122
Q56 What can happen when a permit-to-work system is not effective?					☑	☑		124

PLANNING & IMPLEMENTATION	Communication & training	Contractors	Definitions	Employees	Management	Systems & procedures	Standards	Page
Q57 What is a bridging document?			☑		☑	☑		126
Q58 What is in a bridging document?					☑	☑		128
Q59 What is management of change?			☑		☑	☑		129
Q60 What is an 'in-kind' change?			☑		☑	☑		131
Q61 What should we consider under management of change?					☑	☑		132
Q62 How should we handle management of change?					☑	☑		133
Q63 How do we plan for emergencies?					☑	☑		135
Q64 How do we plan for fire safety?					☑	☑		138
Q65 How do we prevent fires from occurring?					☑	☑		140
Q66 How do we detect fires and raise the alarm?					☑	☑		144
Q67 How do we handle fire-fighting?					☑	☑		148
Q68 What types of fire-fighting equipment are available?					☑	☑		150
Q69 What means of escape from fire should we plan?					☑	☑		151

PLANNING & IMPLEMENTATION

	Communication & training	Contractors	Definitions	Employees	Management	Systems & procedures	Standards	Page
Q70 What do we need to do for fire emergency plans and training?	☑				☑	☑		154
Q71 What is procurement?			☑		☑	☑		157
Q72 What needs to be covered under procurement?					☑	☑		158
Q73 How do we control our contractors?		☑			☑	☑		160
Q74 How do we evaluate and select contractors?		☑			☑	☑		162
Q75 How do we plan contractor work prior to commencement?		☑			☑	☑		164
Q76 How do we train and instruct contractors prior to work commencing?		☑			☑	☑		165
Q77 How do we manage and supervise contractors?		☑			☑	☑		167

EVALUATION

	Communication & training	Contractors	Definitions	Employees	Management	Systems & procedures	Standards	Page
Q78 What are leading indicators?			☑			☑		170
Q79 What are lagging indicators?			☑			☑		172

EVALUATION	Communication & training	Contractors	Definitions	Employees	Management	Systems & procedures	Standards	Page
Q80 How do leading and lagging indicators work?					☑	☑		174
Q81 How do we use leading indicators effectively?					☑	☑		175
Q82 How do we collect leading indicator information?					☑	☑		177
Q83 What is occupational safety and health performance?			☑			☑		180
Q84 Why should we measure occupational safety and health performance?					☑	☑		181
Q85 Why should we analyse safety statistics?					☑	☑		183
Q86 Why do accidents happen?					☑	☑		185
Q87 What human errors are contributory factors to accidents?					☑	☑		188
Q88 What workplace factors are contributory factors to accidents?					☑	☑		190
Q89 Why do we investigate accidents?					☑	☑		193
Q90 How do we investigate accidents?					☑	☑		195
Q91 How do we prevent accidents?					☑	☑		198
Q92 What is an audit?			☑		☑	☑		200
Q93 What do we audit against?					☑	☑		202

EVALUATION	Communication & training	Contractors	Definitions	Employees	Management	Systems & procedures	Standards	Page
Q94 How thorough do we need to be in an audit?					☑	☑		**206**
Q95 How often do we need to audit?					☑	☑		**208**
Q96 What is a management review?			☑		☑	☑		**209**

ACTION FOR IMPROVEMENT	Communication & training	Contractors	Definitions	Employees	Management	Systems & procedures	Standards	Page
Q97 What are corrective actions?			☑			☑		**212**
Q98 How do we manage corrective actions effectively?					☑	☑		**214**
Q99 What is continual improvement?			☑		☑	☑		**217**
Q100 Why is continual improvement important?					☑	☑		**219**

POLICY

Q1   What is health and safety?

'Health' and 'safety' are familiar terms, and most business people know what they are and how to measure them. For this book, within the context of safety management, health is defined as 'occupational health', the health of people in the workplace such that *"people leave their workplace no less healthy than when they arrived"*.

Safety is variously defined as 'secure from harm', 'free from danger' and 'no longer dangerous'. Since safety never can be absolute, for **QUICK WIN SAFETY MANAGEMENT**, it is defined as *"the reduction of risk to a level that is as low as reasonably practicable"*.

Taking these definitions for health and safety into consideration when implementing and operating a safety management system (SMS) should allow management to meet the moral, legal and financial obligations that companies have under current health and safety legislation.

See also

Q2   What is a safety management system?
Q3   How is a safety management system organised?
Q5   What are some of the obstacles to safety management?
Q6   What are the most common hazards in the workplace?
Q15   Who is responsible for safety in an organisation?
Q86   Why do accidents happen?

Q2   What is a safety management system?

An effective safety management system is a documented process for the *"protection of workers from hazards and the elimination of work-related injuries, ill health, diseases, incidents and deaths"*.[1]

For any organisation, a SMS should include:

- The development of a safety policy that outlines the organisation's aims and aspirations.
- Setting goals and objectives across all management levels to improve safety performance.
- Developing a risk assessment process to identify, assess and control hazards.
- Ensuring appropriate training and competency levels appropriate for the size and nature of the business.
- A system for reporting accidents, incidents and other non-conformities within the organisation, and for the analysis and implementation of corrective actions required to prevent their recurrence.
- Documented procedures for all work tasks and the development of permitted systems of work as required.
- Audit and review process for the ongoing assessment of the safety management system.

QUICK WIN SAFETY MANAGEMENT uses the internationally-recognised safety management model based on the International Labour Organization document, *Guidelines on Occupational Safety and Health Management Systems - ILO-OSH 2001*. However, it is important to remember that *ILO-OSH 2001* is not legally binding and is not intended to replace national laws, regulations and accepted standards in your jurisdiction(s). In general, legislation does not prescribe the management systems that employers should implement to comply with their OSH

---

[1]   *ILO-OSH 2001*, Section 1 Objectives.

requirements but, as a minimum, your safety management system must meet the legal occupational safety and health law that applies to your activities at all of your locations.

See also

Q3 How is a safety management system organised?

The *ILO-OSH 2001* system, as shown below, comprises three broad areas:

- The five central sections represent the main components of our safety management system (SMS). Using a defined clockwise workflow, these sections interact and apply across all levels of management and the workforce.

- The inner core, labelled 'Audit', indicates that the use of the audit process across all of the components is essential to ensure that the SMS is working as defined.

- The outer ring, labelled 'Continual Improvement', is an essential ongoing process that must take place in order for the system to improve and evolve once the SMS is operational.

Fig 1 – Main elements of the *ILO-OSH 2001* system

The vast majority of SMS models cover the following sections in much the same order as the ILO model, although there may be differences in terminology and structure. You need to consider:

1 - Policy

- The key principles, aims and objectives of your organisation.
- Incorporating industry-specific best practice requirements.
- Incorporating other management systems (such as quality or environmental management systems) into your SMS.

2 - Organisation

- Defining management structures and levels (by department and by management levels).
- Defining roles and responsibilities.
- Defining reporting hierarchies.

3 - Planning and Implementation

- How your organisation identifies current applicable national laws and regulations, national guidelines, voluntary programmes and other requirements on which the system will be based.
- The identification of workplace hazards and the assessment and management of risk.
- An initial assessment of your organisation's existing safety management structure (if it exists).
- The setting, assessment and measurement of OSH objectives.
- The development of documented procedures for all work activities, including permit-to-work activities where applicable.
- The development of documented management procedures for all safety management activities (such as management of change, management of contractors, emergency response, accident reporting and investigation, etc).

4 - Evaluation

- How your organisation measures safety performance against pre-defined criteria (safety objectives, industry or national standard statistics, etc) and provides feedback into the system.

- How the OSH system is audited and assessed at all levels.

- The measurement of pro-active reports (that indicate there may be undesirable activities happening in the system, such as near-miss reporting, defect reporting, etc) and re-active reports (that document undesirable actives that have happened, such as accident statistics, incident and accident reports, etc).

- How management periodically reviews the OSH performance of the organisation.

5 - Action for Improvement

- The identification of non-conformities within the OSH system and how these are managed.

- The assessment of all OSH activities and the identification of areas for continual improvement.

These five areas make up the core sections in **QUICK WIN SAFETY MANAGEMENT**. Cross-references to the relevant sections within the *ILO-OSH 2001 Guidelines* are indicated in each question thus:

ILO-OSH 2001	3.1. Occupational safety and health policy

See also

Q2 What is a safety management system?
Q4 What is the background to the *ILO-OSH 2001* safety management system?

Q4   What is the background to the *ILO-OSH 2001* safety management system?

The International Labour Organization (ILO) is a United Nations (UN) specialised agency, formed in 1919 as a result of the Treaty of Versailles following the end of World War I, with the aim of dealing with labour issues internationally. Its structure is somewhat unusual for a UN agency in that it has a tripartite governing structure, representing governments, employers and workers. Membership consists of nation-states, of which there are currently over 180. The ILO mission and objectives state that: *"... the ILO is dedicated to bringing decent work and livelihoods, job-related security and better living standards to the people of both poor and rich countries. It helps to attain those goals by promoting rights at work, encouraging opportunities for decent employment, enhancing social protection and strengthening dialogue on work-related issues".*[2]

The motivation for the *ILO-OSH 2001 Guidelines* is best summarised by the document itself: *"The Guidelines were prepared on the basis of a broad-based approach involving the ILO and its tripartite constituents and other stakeholders. They have also been shaped by internationally agreed occupational safety and health principles as defined in relevant international labour standards. Consequently, they provide a unique and powerful instrument for the development of a sustainable safety culture within enterprises and beyond. Workers, organizations, safety and health systems and the environment all stand to benefit".*

Due to its global reach, through its own offices and that of the UN, the ILO has been able to promote its *OSH Guidelines* internationally, with the aim of providing a foundation within which organisations can develop *"coherent policies to protect workers from occupational hazards and risks while improving productivity".*[3]

---

[2]   http://www.ilo.org/global/About_the_ILO/Mission_and_objectives/ lang--en/index.htm.

[3]   *http://www.ilo.org/global/What_we_do/Publications/ILOBookstore/Orderonline/ Books/lang--en/WCMS_PUBL_9221116344_EN/index.htm.*

At a business level, the *OSH Guidelines* are designed to encourage organisations to incorporate safety management into their existing corporate structures as another normal management function and to develop polices and arrangements with the aim of improving OSH performance. If properly implemented, the guidelines provide the means to achieve the now universal legal duty for employers to protect the health and safety of their workers. Thus, the *ILO-OSH 2001 Guidelines* are a good place to start in developing a SMS.

See also

Q3   How is a safety management system organised?

Q5   What are some of the obstacles to safety management?

A 2009 report from the UK's Risk and Regulation Advisory Council (RRAC) highlighted what many small business owners already know: time and money is being wasted in trying to manage their health and safety issues due to confusion about legislation (from mixed messages from Government, legal, commercial bodies and the media) and from a lack of resources. The report showed that many small and medium-sized enterprises (SMEs) were uncertain of their obligations and, therefore, wasted time and money in becoming either over-compliant, implementing inappropriate risk management arrangements or even doing nothing at all.

'Risk-mongers' spreading the wrong message
People who are motivated by self-interest or who are ignorant of health and safety can spread the wrong message. The report gives the example of some safety consultants who *"can inflate the perception of a risk, leading to higher, unnecessary costs and further business for the consultant"*.[4] The print media also, on numerous occasions, has published unfounded or poorly researched stories about supposed health and safety issues that were either inaccurate or non-existent. To counter this almost endemic problem of mis-information on health and safety issues, the UK Health and Safety Executive has sections of its website dedicated to the 'Safety Myth of the Month'[5] and 'Putting the Record Straight'.[6] These sections pick out a news headline or safety myth where a health and safety issue has been incorrectly reported and responds with the correct information.

---

[4]   Risk and Regulation Advisory Council (2009). *Health and safety in small organisations - Reducing uncertainty, building confidence, improving outcomes.*

[5]   http://www.hse.gov.uk/myth/index.htm.

[6]   http://www.hse.gov.uk/press/record.htm.

Contradictory messages

SMEs have many sources to tap into for advice on health and safety, but some advice can come with an agenda, a particular point of view or perspective. Insurers, print and TV media, Government agencies, small business associations and safety consultants are all in the mix on this issue. The end result is confusing and inconsistent messages going out to businesses – the RRAC reported that *"over 50% of the small organisations we questioned had received conflicting messages about health and safety"*.

Lack of SME resources

Health and safety legislation has been developed on the risk assessment model, which provides flexibility for SMEs to tailor risk management controls and solutions to match their specific requirements; however, this flexibility also can cause paralysis in some SMEs where the in-house competence or confidence to apply risk management practices properly is lacking. In reality, risk management is not a 'black art' and the principles are simple to understand and apply for the vast majority of companies.

Sometimes, it can be difficult for SMEs to see a clear path ahead when dealing with health and safety issues but it is important that some effort is put towards this goal, not just for legal compliance reasons but also for the financial benefits of reduced costs. There is no doubt that effective safety management is a sound investment for future business needs.

See also

Q1   What is health and safety?
Q56   What can happen when a permit-to-work system is not effective?
Q62   How should we handle management of change?

Q6   What are the most common hazards in the workplace?

Below is a list of common hazards, which can be low or high risk depending on the circumstances:

- Access and egress, alcohol abuse, biological agents, buildings and facilities, bullying, confined spaces, cash-in-transit activities, company vehicles.

- Display screen equipment, drug abuse, electrical equipment and systems, environment (lighting, etc.), explosive atmospheres, fire, food preparation.

- Hot work, hazardous substances.

- Lifting operations.

- Manual handling, occupational noise exposure.

- Pressurised systems, radiation (ionising and non-ionising).

- Slips, trips and falls.

- Vibration, working at height, workplace vehicles, workplace violence, work equipment.

Note that a hazard is something that has the potential to cause harm, such as a speeding car or a shotgun, while risk is a combination of severity (how severe will be the effects in an incident) and likelihood (the probability of an incident happening). So, on a busy shopping street, a speeding car presents a high risk of an incident (severe effects, high likelihood), while on a race track, it presents a lower risk (less severe effects, lower likelihood). Similarly, a shotgun in the hands of a bank robber is definitely high risk but, in the hands of a professional skeet shooter, the risk of an incident is lower.

See also

Q1   What is health and safety?

Q7   What is a policy statement?

A policy statement is a corporate plan of action that serves to illustrate management's commitment to achieving an objective, such as a safe place of work for a safety policy, zero emissions for an environmental policy, etc. It should state how your organisation plans to go about its activities, what the expectations of senior management are and how, in general terms, you plan to achieve these goals. Remember that policy statements should be specific to your workplace, not a cut-and-paste exercise from another organisation.

When developing policy statements for your organisation, consider:

- Legal requirements to be met.
- Specific requirements of your own organisation.
- Industry standards and best practice requirements that may apply and can be integrated into your policy statement.
- Whether resources are available to ensure that requirements set out in the policy statement can be fulfilled.

Policy statements also should be:

- Concise, well-written, dated and signed by the most senior accountable person(s) in the organisation.
- Communicated to the workforce and be readily accessible.
- Reviewed periodically to ensure that they remain current.

ILO-OSH 2001	3.1. Occupational safety and health policy

See also

Q2	What is a safety management system?
Q8	What policy statements do we need?
Q9	What is the process for writing policy statements?
Q10	Who should write our safety management policy statements?

Q11 What does an occupational safety and health policy look like?

Q12 How do we communicate our occupational safety and health policy statement?

Q15 Who is responsible for safety in an organisation?

Q16 How do we define responsibilities and accountabilities?

Q28 What information needs to be communicated within a safety management system and how?

Q8   What policy statements do we need?

What policy statements are required for your organisation depends on:
- Legal requirements.
- The size and complexity of your organisation's work activities.
- What your organisation is trying to achieve.

An occupational safety and health (OSH) policy statement should be a starting position, although you may need to develop additional policy statements that reflect your organisation's other activities and obligations. For example, since environmental issues are closely tied to OSH and are clearly defined within legislation, if you are dealing with environmental issues, you should develop a policy statement to show that you recognise your obligations and can show how these are managed. Additional policy statements can be developed for other OSH issues such as:
- Quality.
- Environment.
- Waste management.
- Drug and alcohol abuse.
- Anti-bullying.
- No-smoking.
- Right to stop work.

ILO-OSH 2001	3.1. Occupational safety and health policy

See also

Q7   What is a policy statement?
Q9   What is the process for writing policy statements?
Q10   Who should write our safety management policy statements?
Q11   What does an occupational safety and health policy look like?

Q12   How do we communicate our occupational safety and health policy
          statement?

Q13   What should be included in additional policy statements?

Q9   What is the process for writing policy statements?

Policies, like all corporate level documents, need careful consideration during development to ensure that they do not conflict with laws and other formal guidance within your operating jurisdiction(s). It is recommended that, prior to publication, policies are verified by legal experts to ensure compliance.

The following steps illustrate an approach to developing your own compliant policy documents at corporate level. Obviously, the size and complexity of your operations and also the legislative requirement determine which policies you need and how you approach the policy development process. Try this process for an easy way to start:

- Create a team: Bring together the appropriate resources from management, employees and employee representatives to start the policy development process.
- Identify needs: Review the legal requirements for your jurisdiction and identify policies relevant your activities and operations.
- Develop draft and final document versions: Once you have identified your requirements, create draft documents that can go through a review and approval / consultation process before final versions are completed.
- Develop management procedures: It may also be necessary to develop accompanying management procedures that detail how a policy is to be implemented. For example, if you draft an alcohol abuse policy, you may need management procedures to outline:
  - The legal requirements, including right to privacy and confidentiality.
  - Whether alcohol or urine samples can be taken and, if so, under what circumstances.
  - The chain of custody requirements for samples.
  - The requirements and standards for laboratory testing.

- The minimum training levels of persons permitted to take samples.
- The use of kits or single-use testing devices.
- Disciplinary procedures.
- Rehabilitation options.

  At a minimum, policy statements must reflect your legal obligations and requirements, so be very careful that all of your obligations are known and that your supporting management procedures are legal and accurate.

- Publish and communicate: Once a final version of a policy document has been agreed, communicate it throughout your organisation through whatever channels are appropriate.

- Monitor and review: Once the policy has been published, it provides a framework of how your organisation says it conducts itself in the matters relating to the policy statement and can be used to check that policy requirements are being met as stated. Ensure that policy statements are reviewed periodically, usually at two-year intervals.

ILO-OSH 2001	3.1. Occupational safety and health policy

See also

Q7	What is a policy statement?
Q8	What policy statements do we need?
Q10	Who should write our safety management policy statements?
Q11	What does an occupational safety and health policy look like?
Q12	How do we communicate our occupational safety and health policy statement?

Q10 Who should write our safety management policy statements?

Since a policy statement outlines what your organisation sees as its obligations and priorities, it is very important that this 'foundation stone' for your safety management system is written by senior management (managing director / chief executive officer or equivalent) and endorsed by the Board (where appropriate). It is also vital that these key personnel show their commitment to effective safety management both by signing and dating safety-related policy statements and by reviewing them periodically (usually at least every two years).

If your organisation has a senior management member who is specifically assigned responsibility for occupational safety and health issues, that person should sign the occupational health and safety policy statement too.

ILO-OSH 2001	3.1. Occupational safety and health policy

See also

Q7 What is a policy statement?
Q9 What is the process for writing policy statements?
Q11 What does an occupational safety and health policy look like?

Q11   What does an occupational safety and health policy look like?

The occupational safety and health (OSH) policy statement should define key safety management principles and objectives that the organisation subscribes to.

Sample health and safety policy statement
The following is an example of an OSH policy statement:

> *It is the policy of [insert company name] to comply fully with the requirements of the [insert name and date of Act] and to ensure so far as is reasonably practicable the safety, health and welfare of all employees at our place of work.*
>
> *In pursuance of the general statement of the safety policy, the company shall:*

- *Provide and maintain a safe place of work and safe systems of work.*
- *Provide safe means of access and egress.*
- *Provide and maintain safe work equipment.*
- *Carry out risk assessments and implement control measures as required.*
- *Protect, so far as is reasonably practicable, persons not employed by this company who may be affected by our activities.*
- *Consult employees on matters of health and safety.*
- *Promote the reporting of non-compliance in behaviours, equipment, processes and procedures.*
- *Provide such information, training and supervision as may be required to work safely.*
- *Prepare and periodically review emergency plans.*
- *Designate staff with emergency duties.*
- *Provide and maintain adequate welfare facilities.*

- *Provide a competent person to assist in securing the health and safety of employees.*

- *Provide sufficient resources to achieve the Company's safety objectives.*

- *Review and audit our safety management system to ensure continual improvement.*

Employees also have clearly defined responsibilities under the Act to co-operate with management to achieve a safe place of work and to take reasonable care of themselves and others.

This policy will be kept up-to-date, particularly as the business changes in nature and size. To ensure this, the policy shall be reviewed annually.

ILO-OSH 2001	3.1. Occupational safety and health policy

See also

Q7   What is a policy statement?
Q9   What is the process for writing policy statements?
Q10 Who should write our safety management policy statements?
Q12 How do we communicate our occupational safety and health policy statement?

Q12   How do we communicate our occupational safety and health policy statement?

Since your occupational safety and health (OSH) policy statement is a management pledge and requires the total involvement of everyone within your organisation, it must be communicated effectively to everyone internally – and externally, too, to contractors and suppliers as appropriate.

However, the most important type of OSH policy communication is the workforce seeing management at all levels applying the policy to their own activities, and thus by example showing that it is taken seriously. When staff are told to do something, but then see that managers not doing it themselves, all the policy statements, emails and paper in the world cannot overcome this obstacle. Good personal example to others is top of the list for effective communication.

Here are some other ways of communicating your OSH policy statement to the people who are required to see it, which will help it to become embedded into your day-to-day business activities:

- A positive personal example set by all levels of management during day-to-day operations that endorse the policy philosophy.

- Posting policy statements on notice boards, reception areas and in other busy locations.

- Including the OSH policy as part of your organisation's safety induction training for new employees.

- Tailoring job descriptions to ensure that the duties (including responsibility and accountability) of staff reflect the OSH policy statement's aims.

- Management involvement in safety committees, general safety meetings, and audits and inspections (where appropriate).

- Promotion of safety representatives within your organisation so that policy aims are known at the shop-floor level.

- Use of newsletters and internal memos from senior management to promote OSH policy, OSH issues and to re-inforce the company's safety philosophy.

| *ILO-OSH 2001* | 3.1. Occupational safety and health policy |

See also

Q7   What is a policy statement?
Q27 What is communication?
Q28 What information needs to be communicated within a safety management system and how?

Q13   What should be included in additional policy statements?

Additional secondary policy statements can be developed for occupational safety and health issues. Examples can include:

- Drug and alcohol abuse.
- Anti-bullying.
- No smoking.
- Driving company vehicles.

Developing policy statements depends on what activities your organisation undertakes, what regulations apply to it and what the aims and targets of senior management are. Some of these policy statements will need to take into consideration local laws, regulations and formal guidance issued from a competent authority.

Below are some safety-related policy areas and guidance on the issues to be considered in your policy statements and explanatory management procedures. The lists are not exhaustive so make sure that you do your own homework on these policies.

Activity	Outline content to be considered
Drug / Drug and alcohol abuse	Legal framework – the legal requirements, including right to privacy and confidentiality. Circumstances where testing can be conducted. Testing – whether alcohol or urine samples can be taken and under what circumstances. Chain of custody – requirements for samples. Standards - requirements and standards for laboratory testing. Competency – minimum training levels of persons permitted to take samples. Disciplinary procedures – what are the options if drug /alcohol abuse is detected. Right of appeal. Rehabilitation options.

Activity	Outline content to be considered
Anti-Bullying	Legal framework – the legal requirements that apply to your organisation. Company statement – for example: *"the aims of the organisation are not to tolerate bullying, employees' responsibility to maintain a work environment free from bullying, etc"*. Definition – define what bullying is (use a legal definition if available). Dealing with allegations of bullying (informally) – define the system for handling bullying complaints informally within your organisation, including the complaint, intervention and closure. Dealing with allegations of bullying (formally) – define the system for handling bullying complaints formally within your organisation, including complaints, investigation, findings, discipline procedure and closure.
No Smoking	Legal framework – the legal requirements that apply to your organisation. Company statement – for example, *"the aims of the organisation for the promotion of good health, to have a healthy workforce, etc"*. Common and work locations – outline the no-smoking areas at your facility, including company vehicles if required. Highlight any dangerous or restricted areas where smoking could be a fire or explosion hazard. Smoking areas – outline the areas where smoking is permitted. Housekeeping – outline provision for the safe disposal of cigarette waste (metal bins, sand bins, etc.) Health promotion – outline any schemes in place to supply nicotine patches, gums or other incentives to stop smoking. Disciplinary procedures – what are the options if smoking is detected outside of permitted areas.

Activity	Outline content to be considered
Driving company vehicles	Legal framework – the legal requirements that apply to your organisation. Company statement – for example, *"the aims of the organisation for the reduction of accidents in company vehicles, better standards of driving within the company, better management of emergencies situations, etc".* Standards – specify minimum standards for company vehicles, including servicing, periodic /daily checks, system for reporting problems, minimum safety equipment to be carried in the vehicle (first aid kits, hi-visibility vests, warning triangles), etc. Emergency – specify emergency contacts (including out-of-hours contacts), provide contact information for insurance and breakdown services. Journey management – selection of routes for journeys (such that journeys can be achieved without speeding), taking breaks during driving, maximum working hours including driving, consideration for weather conditions, driving abroad, etc. Phone use – use of mobile phones in cars, supply of hands-free kits, etc. Avoiding driving – alternate travel arrangements (train, bus, etc), use of phone and video conference calls. Training – consider specialist driver training courses for drivers who exceed a certain mileage annually.

ILO-OSH 2001	3.1. Occupational safety and health policy

See also

Q7   What is a policy statement?

Q8   What policy statements do we need?

Q14   What is meant by worker participation?

Worker participation regarding health and safety is increasingly important in the workplace. In fact, it is now considered to be so critical in the general social model in the EU that it is defined as a fundamental right in the European Charter of Fundamental Rights.[7] In terms of safety management, worker participation can be defined as: *"providing the opportunity for employees to make a positive contribution to improve occupational safety and health in the workplace"*.

For employers developing or running an already-established safety management system (SMS), in practice this means:

- Recognising the often legally-defined role of the workers' safety and health representatives.
- Ensuring that workers are consulted, informed and trained on all aspects of occupational safety and health (OSH), including emergency arrangements, associated with their work.
- Making arrangements for workers to have the time / resources to participate actively in the processes defined within the OSH system.
- Consulting with the workforce through a variety of media to ensure a frequent exchange of views and information.

A SMS must define a wide range of processes, activities and procedures to support worker participation. These may include:

- Workers assisting in the development of work procedures, work instructions and workplace risk assessments.
- Encouraging workers to report unsafe acts, unsafe conditions, suggestions for improvements, etc, through a documented reporting system.
- Providing workers with training and instruction appropriate to the work tasks.

---

[7]   2000/C 364/01, *Official Journal of the European Communities, Charter of Fundamental Rights of the European Union* – Document C364 18.12.2000.

- Attendance at safety meetings, safety committees and other related meetings (where OSH may be on the agenda).
- Workers and management taking part in audits and inspections of the workplace.
- Promotion of safety representatives within the workforce.
- Safety incentive schemes to encourage worker participation.
- Clearly-defined job descriptions, including employee responsibilities as defined in national OSH legislation.
- Documented 'Stop the Job' or 'Right to Refuse Work' policies.

In the real world of course, negative influences (either individually or collectively) can impact severely on the extent to which workers are able, or allowed, to get involved in OSH matters at work, regardless of the legal framework in their jurisdiction. These include:

- A difficult economic climate, which can impact on employment prospects and, therefore, on morale and motivation at work.
- Management with a lack of commitment to health and safety (normally resulting in a poor safety culture).
- Safety representatives or delegates who are chosen by management and not by the workforce.
- Lack of information available in the workplace about the legal duties and responsibilities of employer and employees regarding health and safety.

ILO-OSH 2001	3.2. Policy / Worker Participation

See also

Q15   Who is responsible for safety in an organisation?
Q21   How do we identify our safety training requirements?
Q24   How do we format a work procedure?
Q41   What is risk assessment?
Q92   What is an audit?

ORGANISATION

Q15   Who is responsible for safety in an organisation?

In general terms, national occupational safety and health (OSH) statute law (and legal precedents set under common law) places the responsibility for the health, safety and welfare of employees and other third parties on employers (the term 'employers' may relate to directors, boards of management or other defined legal entities, depending on the legal structure of the organisation). While it is recognised that employees also have defined duties under legislation, in terms of safety management which requires the leadership, commitment and resources to implement, in **QUICK WIN SAFETY MANAGEMENT** responsibility is viewed from the employer perspective.

The appointment of directors at board level with direct responsibility for OSH is now seen as a norm in many organisations, ensuring strong leadership, focus and consistency.[8]

In addition, many organisations now consider OSH in the context of corporate governance and integrate safety risk along with those more established business risk functions such as finance, recognising that significant lapses in OSH can have a substantial impact on reputation and a potential significant financial impact, too (for example, fines under the UK's *Corporate Manslaughter and Corporate Homicide Act, 2007* are unlimited).

Various national OSH statute laws define the responsibilities of employers in different terms, but generally the aim is the same: *"... that every employer shall ensure, so far as is reasonably practicable, the safety, health and welfare at work of their employees"*.

The Irish *Safety, Health and Welfare at Work Act, 2005* is typical of OSH legislation. Under this Act, the duties of employers can be summarised as:

---

[8]   Health and Safety Executive (2006). *Health and Safety Responsibilities of Company Directors and Management Board Members: 2001, 2003 and 2005 surveys, Final Report*, HSE Research Report 414, p.iii.

- Managing and conducting work activities to ensure the safety[9] at work of employees and to prevent any improper conduct or behaviour likely to put the safety of employees at risk.
- The design, provision and maintenance of the place of work in a condition that is safe and without risk to health, has safe means of access to and egress from it, and that plant and machinery or any other articles are safe and without risk to health.
- Ensuring the safety and the prevention of risk to health at work of employees relating to the use of any article, substance, radiations and physical agents.
- Providing systems of work that are planned, organised, performed, maintained and revised as appropriate.
- Providing and maintaining facilities and arrangements for employee welfare.
- Providing the information, instruction, training and supervision necessary to ensure the safety of employees.
- Determining and implementing measures for the protection of the safety of employees when identifying hazards and carrying out a risk assessment and ensuring that the measures take account of changing circumstances and the general principles of prevention.
- Where risks cannot be eliminated or adequately controlled, providing and maintaining suitable protective clothing and equipment to ensure the safety of employees.
- Preparing and revising adequate plans and procedures to be followed and measures to be taken in the case of an emergency or serious and imminent danger.
- Reporting accidents and dangerous occurrences.
- Obtaining, where necessary, the services of a competent person to ensure the safety of employees.

---

[9] In this list of duties, the term 'safety' can be understood to mean safety, health and welfare.

Since employer responsibilities are well-defined and wide-ranging, some form of safety management system, such as the *ILO-OSH 2001* model, needs to be in place to organise and manage these, as well as to manage risk in the workplace.

ILO-OSH 2001	3.3. Responsibility and Accountability

See also

Q1   What is health and safety?

Q3   How is a safety management system organised?

Q16   How do we define responsibilities and accountabilities?

Responsibility can be defined as: *"the duty or obligation to carry out a task or role and for which the person has been provided with the necessary information, instruction, training and supervision, but which can have consequences and penalties for failure"*.

Accountability can be defined as: *"the acknowledgment and assumption of responsibility for actions and decisions made within the defined scope of a documented job description, including the obligation to report, explain and be answerable for any possible negative consequences"*.

Responsibility and accountability are fundamental to the foundation of an effective safety management system (SMS) and should not exist without each other. It is not possible to have effective line management where responsibilities are not clearly defined; how could an employee be held accountable for an act or omission in the event of an accident or incident if their responsibilities are not defined? Organisations that do not define clearly the roles and responsibilities of their employees and managers risk breaking the law – both in terms of occupational safety and health (OSH) law and also employment law, depending on the requirements in your jurisdiction(s).

The easiest approach to defining accountabilities and responsibilities is to create job descriptions for all of the roles within your organisation, from the chief executive down to shop-floor workers, taking the following into consideration:

- Job descriptions must comply with the minimum legal requirements as outlined in employment and OSH law within your jurisdiction(s).
- Whoever has responsibility for defining job descriptions within your organisation must understand the specific occupational health and safety duties that apply to your organisation.

- Job descriptions must be defined, documented and communicated to employees, not just when hiring people, but also where there may be changes to employment or OSH law.
- Job descriptions should be made available to all of your employees within your SMS or human resources system.
- The format should be clear, concise and uniform across the organisation and cover OSH requirements.

ILO-OSH 2001	3.3. Responsibility and Accountability

See also

Q3   How is a safety management system organised?
Q17 What does a job description look like?

Q17   What does a job description look like?

The following template for job descriptions is just one of the many examples available for organisations to use, but this one specifically addresses health and safety issues and reporting requirements.

A job description should contain the following information:

- Job title: Title as defined within the organisation, such as 'Warehouse Manager' or 'Receptionist'.
- Location: The actual geographical location where the person will work, such as 'Oslo head office' or 'Dublin Transport Depot – Warehouse A'.
- Grade / level: Where there is a requirement to create different grades or levels within a defined job title, such as 'Administrative Assistant – Level 2' or 'HSE Advisor – Grade 1'.
- Reporting to: The role or position that the employee reports to when in this role – perhaps more than one person, depending on specific requirements.
- Duties and responsibilities: Outline what are considered to be the basic tasks assigned for this role.
- Health and safety duties: Specify any health and safety duties where they may apply, such as 'Weekly inspections of the food preparation areas according to the HACCP plan' or 'Fire Warden – Ground Floor', although these HSE roles will need to be defined in more detail within the safety management system. Employees must be fully trained for the roles to which they are assigned.
- Qualifications: The basic academic and vocational qualifications required for the role, such as 'Degree / Diploma-qualified in Electronic Engineering, Computer Science ' or 'Chartered Member of Institute of Occupational Safety and Health'.
- Experience: Outline the relevant experience for the role.
- Working conditions: Outline relevant working conditions, such as working hours, travel requirements etc.

- Additional requirements: Any additional comments as required.

The following is an example of a possible job description for the role of a Fork Lift Truck Driver in a storage warehouse.

Job Title	Fork Lift Truck Operator.
Location	Distribution and Storage Facility – Warehouses A and B.
Grade / Level	Grade 3.
Reporting To	Storage Warehouse Supervisor.
Duties and Responsibilities	Load and unload HGV vehicles. Palletise product ready for loading and cling-wrap finished product as required. Maintain a clean, safe and tour-ready facility. Assist in warehouse stock-keeping as required. Responsible for re-shelving and re-stocking product. Other related duties as required.
HSE Duties	Carry out all forklift safety checks and inspections as defined in Company procedures. Actively participate in safety programs. Wearing of appropriate PPE as specified. Comply with all safety duties defined in Company Handbook (Chapter 3 – Health and safety responsibilities).
Qualifications	Minimum 5 'O' Levels of grade C and above. In-date RTITB operator certification for counterbalance and pivot steer (Bendi/Flexi) vehicles.
Experience	Minimum 1 year in similar role.
Working Conditions	Full-time. Working hours – 08:00 to 17:30 hours (with breaks). 5 day week – no weekend work.
Additional Requirements	Manual Handling Training.

See also

Q16   How do we define responsibilities and accountabilities?

Q18　Why is training an important part of a safety management system?

Training can be defined as: *"activities designed to facilitate the learning and development of new and existing skills, and to improve the performance of specific tasks or roles"*.

The provision of training within an organisation's safety management system (SMS) is important because:

- It meets an employer's legal duty to protect the health and safety of its employees.
- It helps to identify and reduce hazards in the workplace.
- It helps to reduce the risk of workplace accidents and, therefore, reduces the cost of workplace accidents.
- It assists in developing a health and safety culture within the organisation.
- It reduces the likelihood of negligence claims, where the issue of competence may be brought up in legal actions against the organisation.

Although training has an important role to play in improving the levels of occupational safety and health (OSH) performance, it will only be effective as part of a holistic SMS. For example, if through the risk assessment process you identify that you need to have training for workers for entry into confined spaces and then carry out that accredited training, but do not support it with a permit-to-work system, documented isolation procedures, the correct personal protective equipment (PPE) and respiratory protective equipment (RPE), no supervision or emergency planning, has the risk of confined space entry work been reduced as low as reasonably practicable? Clearly, 'No'. But adding these other safety management elements mentioned above to appropriate training increases confidence that your organisation has a robust system to manage the risk effectively.

Of course, training is not only beneficial to the organisation but also can have significant benefits to individuals, including:

- Improved confidence and morale in carrying out work tasks.
- Reduction in the probability of an accident or incident involving the individual.
- Providing accredited qualifications to workers improves their worth to the organisation and may result in improved pay and conditions.
- Improving the skill set of workers.
- A better understanding of both the employer's and individuals' responsibilities in the workplace for health and safety.

ILO-OSH 2001	3.4.3. Competence and Training

See also

Q3   How is a safety management system organised?
Q20   How do we develop safety training?
Q21   How do we identify our safety training requirements?

Q19   How do we define competence?

The word 'competence' has been defined variously as: *"the application of skills and knowledge to effective practice expectations in the workplace"*;[10] *"a person's ability to perform to a satisfactory level in the workplace, including the person's ability to transfer and apply skills and knowledge to new situations, and to achieve agreed outcomes"*;[11] and *"education, work experience and training, or a combination of these"*.[12]

The use of a general definition of competence as stated above is probably sufficient for most low-risk workplaces. However, if employees or contractors of your organisation undertake any specialist or non-standard work activities, you should consider developing your own definition of competence (within job descriptions, for example) for roles where there are:

- No defined competencies in an industry sector or within the law.
- General defined competencies specified by representative or other professional bodies.
- General defined competencies specified within an industry sector but not defined in law.
- Specific requirements defined in law.

You must assess the roles and responsibilities that exist within your organisation and identify how to achieve compliance, first with the law and then with your own defined safety management system (SMS).

Once established, these requirements must be documented within your SMS in job descriptions, communicated to those people to whom they apply and the necessary resources made available to fulfil the competency requirement.

---

[10]   UK Training and Development Agency for Schools (2008). *Glossary.*
[11]   Worksafe Australia. *Occupational Health and Safety and Competency Based Training - Some Questions Answered.*
[12]   *ILO-OSH 2001.*

ILO-OSH 2001	3.4.1. / 3.4.2. Competence and Training

See also

Q16 How do we define responsibilities and accountabilities?
Q18 Why is training an important part of a safety management system?

Q20 How do we develop safety training?

When developing a plan to implement occupational safety and health (OSH) training within your organisation, the following aspects need to be taken into account.

General considerations

- Document the training philosophy and approach that your organisation takes in your safety management system (SMS) management procedures.
- Consider the training requirements across all levels of the organisation's management, from the shop-floor to the boardroom.
- Ensure that sufficient personnel and financial resources are available to support a planned and long-term training strategy.
- Ensure that training providers to your organisation have the required competency and accreditations appropriate to the training provided.
- When training takes place, ensure that there is a process for participants to evaluate and comment on the training provided.
- Ensure that a procedure is in place to verify the training standards and qualifications claimed both by contractors (when tendering for work) and by prospective employees (when applying for employment).

Specific training considerations

Address the following within your SMS for training, with the minimum acceptable standard being the legal requirement that applies to your organisation:

- Ensure suitable training has been identified within your organisation to cover:
  - Normal workplace risks (from your risk assessment process).
  - Specialised workplace risks (such as confined space entry training, etc).

- Safety inductions (including employees, contractors and visitors).
- Emergency procedures and precautions.
- Contract and temporary staff.

- Consider the training requirements of employees who may have particular requirements or specialist needs, such as young people, new recruits or foreign workers.

- Ensure that training is reviewed periodically to ensure its continuing relevance and effectiveness.

- Provide appropriate refresher training where a requirement has been identified.

- Provide OSH training at no cost to the participants and, where appropriate, during normal working hours.

- Ensure that training is provided in a manner, form and language appropriate to your workforce.

- Ensure that your organisation identifies and meets the minimum legal requirement for OSH training for specific roles.

- Ensure that employees are required to attend appropriate training that relates to their work activities (ensure that this requirement is outlined in job descriptions).

- Ensure that attendance and performance at OSH training is taken into consideration during the personnel performance review process.

- Identify the milestones where tailored and appropriate training should be provided, such as:
  - When new employees are hired.
  - When employees are transferred to new work tasks.
  - When new equipment, technologies and systems are introduced into the workplace.

ILO-OSH 2001	3.4.3. Competence and Training

See also

Q21 How do we identify our safety training requirements?
Q41 What is risk assessment?
Q59 What is management of change?
Q63 How do we plan for emergencies?
Q76 How do we train and instruct contractors prior to work commencing?

Q21   How do we identify our safety training requirements?

Before your organisation can develop and implement an occupational safety and health (OSH) training programme, you must take into account a number of critical factors that influence your training needs, including:

- The size of your organisation.
- The jurisdiction(s) your organisation operates in (as a minimum, you must meet the local regulatory requirements in each jurisdiction).
- The scope of work that your organisation is engaged in on a day-to-day basis.
- The complexity of tasks that your organisation is engaged in on a day-to-day basis.
- The management structure and, therefore, the different levels of responsibility assigned within your organisation.
- The approach your organisation takes to develop training across your organisation.

Size of the organisation
Size (in terms of both personnel and financial resources) will dictate whether there is a requirement for dedicated training resources within the Safety or Human Resources (HR) departments to plan, organise, pay for and co-ordinate training. OSH training for larger organisations often is just one of many categories of training required for employees and can dovetail in with vocational and other training requirements.

Jurisdictions
Although OSH regulatory requirements for employers tend to be similar in many parts of the developed world, there can be significant differences when organisations operate in developing or third world environments. As a minimum requirement, every organisation must meet the OSH standard(s) that apply in law at their work locations. Of course, ideally,

the organisation's safety management system should exceed this minimum requirement but, in the event of an accident or incident becoming a legal issue, the question of compliance with the law is always the starting position.

Scope and complexity of operations

The type of OSH training required depends on the hazards and risks within the workplace. A small, low risk environment, such as an administration office, will have minimal training requirements, whereas a construction company that specialises in scaffolding and rope access on offshore oil platforms will have a substantial training requirement.

Management structure

It is important that OSH training covers all levels of management. Although it may appear that workers who carry out day-to-day tasks on the shop-floor are the only ones who require training, this is a mistake. Even senior management may not be aware of their legal requirements and obligations regarding OSH and, often, may not have had up-to-date and relevant instruction about safety management and its potential benefits to an organisation. The type and scope of training may change but the need is universal.

Approach to training

Training must be implemented in a planned and structured way so it is clear what is to be achieved at each stage of the process. Such an approach could include these stages:

- Identification of training needs and requirements.
- Development of the training programme content.
- Implementation of the training programme.
- Evaluation and review process.

In addition, there should be a feedback loop as part of the evaluation and review process. Since the cost of putting in place a training programme can be significant, it is vital that the organisation understands where it is getting value and where changes and improvements need to be made on an ongoing basis.

ILO-OSH 2001	3.4.3. Competence and Training

See also

Q15 Who is responsible for safety in an organisation?
Q20 How do we develop safety training?
Q41 What is risk assessment?
Q63 How do we plan for emergencies?

Q22 How do we control safety management system documents?

Many documents are generated when establishing and implementing a safety management system (SMS), including (but not limited to):

- Policy statements, vision and mission statements and corporate safety objectives.
- Management and work procedures.
- Internal audits and inspections.
- Safety meeting agendas and meeting minutes.
- Permit-to-work forms.
- Risk assessments.
- Corrective actions plans.
- Miscellaneous blank forms and checklists.

With so much documentation within the system, it becomes increasingly important to ensure that all safety management documents are controlled according to defined criteria collectively known as 'document control'. Document control ensures that documentation within an organisation is:

- Clear and unambiguous in its instruction and information.
- Reviewed and approved for use.
- Secure from unauthorised or inadvertent change.
- Accessible and easy-to-use.
- Reliable and accurate when providing communication across the organisation.
- Uniform in its presentation, which promotes a consistent and professional corporate identity.

The basic elements of document control that should be considered are:

- Layout, content and formatting: How are the documents in the system to be formatted? What basic content requirements do we have across the range of documents we produce?

- Document ownership: Who has responsibility for documents in the system? Who has permission to modify and change these? What are the criteria for changes and updates?

- Review and approval process: What is the defined process for adding new documents into the system? What is the defined process for the ongoing review and approval of existing documents?

- Document change requests: Where updates to documents are required due to changes in work practices, changes to equipment or other modifying factors, how are these updates initially requested, rejected or implemented?

- Archiving obsolete documentation: What time period does the organisation specify before a document is archived? What is the archiving method? This is especially important for records generated within the SMS, where there may be a legal requirement to keep certain records for a specified time period or where there could be confidentiality issues (such as health surveillance records, medicals, drug and alcohol test results, etc.), even if these are maintained by a third party.

- Managing external documents used within the organisation: There could be many external documents used within the day-to-day operation of a company, such as external reference documents, equipment test certificates, etc. How are these tracked to ensure that they remain in date and accurate?

For companies using ISO 9001-2008,[13] or considering the use of this quality standard, there are clear guidelines for the 'Control of Documents' within section 4.2.3. The above criteria are for general use but also provide the majority of the quality standard elements too.

---

[13]   BS EN ISO 9001:2008 - Quality Management Systems – Requirements.

ILO-OSH 2001   3.5.2. OSH Management System Documentation

See also

Q23   How do we format corporate safety documents?

The general content issues you need to consider for your organisation's corporate safety documents include:

- Title: A clear, unambiguous title describing the purpose of the document.

- Date: Date the document with a 'valid from' date in an agreed date format (such as DD-MMM-YY). The date format that tends to be used in North America (mmddyyyy) is not recommended for document control as the juxtaposition of the numbers for the day and month can cause unnecessary confusion when used internationally. Instead, use YY-MM-DD, which can assist in auto-indexing where documents are listed within your safety management system and is becoming more commonly used internationally.

- Document ID: A unique reference number to identify each document across an organisation.

- Document Status: To indicate the status of the document (final, draft, for review, etc).

- Page Numbering: To indicate individual page numbers and total number of pages for all pages.

- Miscellaneous Statements: For example, text indicating 'Uncontrolled when printed'.

When the basic document content components have been agreed, the next step is to develop basic formatting criteria:

- Page set-up: Paper size, margins, justification and header / footer size.

- Font: Define font and font size to be used for titles, general text, headers / footers, etc.

- Style: Define the settings for style bullets, numbering, paragraph formatting, etc.

Once these elements have been developed, you have a basic template available to ensure consistent documentation of your safety management system throughout your organisation.

ILO-OSH 2001 | 3.5.2. OSH Management System Documentation

See also

Q22   How do we control safety management system documents?
Q24   How do we format a work procedure?
Q26   How do we document management procedures?

Q24　How do we format a work procedure?

The layout format for a work procedure can be modified to suit your own circumstances but the example below is a good starting point where no format is currently available.

Procedure elements	Description
Purpose and Scope	Outlines the primary purpose of the document and should be short and concise.
Definitions	Any abbreviations, acronyms and terms used in the procedure that might cause confusion and require clarification should be defined here.
Responsibilities	Defines the responsibilities for each person involved in the procedure, to be indicated by job position.
Personal Protective Equipment	Provides specific details of the personal protective equipment (PPE) required for the procedure.
Procedure – Preparatory Actions	Provides specific details on any actions to be carried out before the procedure itself. Including the requirement for a toolbox talk, completing a permit-to-work form, organisation and cleaning of work areas or safety precautions and other appropriate planning or preparatory actions.
Procedure - Process	Provides a specific workflow on the actions to be carried out during the procedure. Actions should be described in the correct sequence of steps, relevant to areas in which they occur and be in sufficient detail so that the purpose of the procedure can be easily understood and followed. The use of explanatory photographs or even embedded video (in more advanced electronic systems) also should be considered.
Records	Provides information on what records are generated from this procedure – for example, a completed permit-to-work form, attendance sheet or completed checklist.

The most suitable format for your own organisation is the one that most closely reflects your own requirements.

| *ILO-OSH 2001* | 3.5.2. OSH Management System Documentation |

See also

Q23 How do we format corporate safety documents?
Q26 How do we document management procedures?

Q25   What are management procedures?

Management procedures are a useful tool to define in general terms the processes and practices that an organisation subscribes to across its operations. They can be defined as: *"corporate documents designed to provide basic information on how an organisation implements specific safety management functions by outlining minimum standards, general guidance and best practice adopted by the organisation which are most appropriate for that specific function"*.

Management procedures sit below corporate documents such as policy statements and vision and mission statements, but above local operational work procedures for which they form the basis.

ILO-OSH 2001	3.5.1. / 3.5.2. OSH Management System Documentation

See also

Q7   What is a policy statement?
Q23   How do we format corporate safety documents?
Q24   How do we format a work procedure?
Q26   How do we document management procedures?

Q26 How do we document management procedures?

If management procedures are adopted, the size and scope of work of your organisation is critical in deciding how these are to be defined and documented within your safety management system (SMS).

In the example below, to keep things simple, only four main headings have been used to define management procedures (with a simple numbering system), but other variations are possible depending on your own organisation's requirements and preferences. The sample comments in italics describe the type of content that could be included in these individual management procedures.

MP-1 Organisation

- MP-101 Policy and Objectives – *Specify the organisation's current policy statements and corporate safety objectives.*

- MP-102 Responsibility – *Reference where job descriptions can be found and what job descriptions should define in detail for all responsibilities and accountabilities, including health and safety duties. Include organisation charts to illustrate the organisational structure at corporate and department levels with clear reporting lines, etc.*

- MP-103 Competence and Training – *Specify minimum competence and training requirements for employees / minimum standards or accreditations for defined safety courses and for safety training providers. Define how training and refresher training will be tracked. Define the requirement and minimum standards for third party competent persons, etc.*

- MP-104 Documentation and Document Control – *Outline how safety management documentation is to be managed. Define responsibility for document management. Detail minimum standards and requirements for the organisation's internal and external safety management documentation, etc.*

- MP-105 Safety Communications – *Define the safety communications regime within the organisation, such as management and employee meeting requirements, safety and health promotion campaigns, safety representatives, new employee safety induction process, safety signage policy and implementation, etc.*

MP-2 Hazard Prevention

- MP-201 Emergency Response – *Outline the organisation's emergency response plans. Define what legal requirements must be complied with. Define responsibilities and training requirements within emergency planning. Outline resources available. Define drills, etc.*

- MP-202 Health and Safety Plans – *Define what plans are used within the organisation, such as project plans, operational plans, etc. Define minimum plan requirements / responsibilities for development and maintenance of plans, etc.*

- MP-203 Personal Protective Equipment (PPE) and Respiratory Protective Equipment (RPE) – *Define requirements for the use of PPE and RPE / standards that equipment must be manufactured to / training, maintenance and inspection requirements, etc.*

- MP-204 Management of Change (MoC) – *Define what processes the MoC procedure applies to and how the MoC process is to be managed. Define responsibilities for MoC, etc.*

- MP-205 Risk Management – *Outline what risk management systems are in place. Define the risk matrix used for quantifying risk. Outline what systems are to be used for assessing risk, such as job safety analysis or other methods. Define hierarchy of controls to be used. Define whether hazard register is used, etc.*

Section MP-2 can be expanded by detailing additional hazard prevention processes that your organisation finds suitable, such as planned preventive maintenance.

MP-3 Safe Work Practices

- MP-301 Permit-to-Work (PTW) – *Define what activities are covered under the permit-to-work system. Outline how the system works. Define responsibilities for PTW. Specify minimum training requirements for persons undertaking PTW activities, etc.*

Section MP-3 can be expanded on by detailing the safe work practices that the organisation carries out and which are relevant to it as a risk reduction measure.

MP-4 Evaluation and Improvement

- MP-401 Accident Investigation – *Define the procedure for accident investigations. Define responsibilities and training requirements / investigation team composition, etc.*

- MP-402 Incident Reporting – *Outline the systems to be used to define what accidents and incidents are reported internally and are legally reportable to the authorities for incident and accident statistics.*

- MP-403 Preventative and Corrective Actions – *Define the system for generating, tracking and managing corrective actions. Define responsibilities for managing corrective actions, etc.*

- MP-404 Audit and Review – *Define the system in place for internal and external auditing and inspections within the organisation. Define system in place for auditing of external contractors, etc.*

ILO-OSH 2001	3.5.1. / 3.5.2. OSH Management System Documentation

See also

Q7	What is a policy statement?
Q15	Who is responsible for safety in an organisation?
Q19	How do we define competence?
Q20	How do we develop safety training?
Q22	How do we control safety management system documents?

Q27   What is communication?

In a business sense, communication can be defined as: *"effectively conveying and receiving clear and unambiguous information in a timely manner and where both parties understand and interpret the information in the same way"*.

Information that is clear, unambiguous and interpreted the same way seems like an obvious statement but, as George Bernard Shaw said, *"the single biggest problem in communication is the illusion that it has taken place"*. So it is vital that within your own organisation's safety management system (SMS), arrangements and procedures should be established and maintained specifically for communicating occupational safety and health (OSH) matters including:

- Receiving, documenting and responding appropriately to internal and external communications.

- Ensuring effective internal communications between relevant levels and functions of management and of the organisation.

- Ensuring that the concerns, ideas and inputs of workers and their representatives are received, considered and responded to.

Communications covers a wide range of media, most of which have some part to play within your organisation for the distribution of information, including:

- Formal and informal management and employee meetings.

- Documentation (such as work procedures, reference documents, corporate polices, safety alerts, manuals, etc).

- Safety and information posters.

- Safety and hazard signs and sign-boards.

- Safety notice-boards.

- Training and instruction.

- Reporting and suggestion processes (reporting defects, leading and lagging indicators, etc).

- Safety inductions and safety tours.
- Emergency drills and exercises.

ILO-OSH 2001	3.6. Communication

See also

Q28 What information needs to be communicated within a safety management system and how?

The appropriateness of the information to be communicated within a safety management system and the methods chosen to do so depends on the local and national legal requirements that apply, the operational size and scope of your organisation and the stated requirements of corporate management policies and objectives.

The table below summarises a typical approach:

	Information	Method
Corporate	Policies and objectives	To be developed by senior management and periodically reviewed. Made available in printed format and posted at key locations.
	Management procedures	To be developed by competent persons and periodically reviewed. Made available in printed and / or electronic format at work locations.
	Safety induction	Tour and review of organisational arrangements for health and safety during new employee company induction. New employees to sign and date safety induction record.
	Job descriptions[14]	Signed and dated printed copy provided to individual employees and record kept by HR department.

---

[14] Including OSH responsibilities and defining any specific emergency duties, etc. Requirements for OSH within job descriptions may be defined in employment legislation in your jurisdiction.

	Information	Method
Operational	Work procedures	To be developed by competent persons and periodically reviewed. Made available in printed and / or electronic format at work locations.
	Safety alerts and updates	To be made available in printed and / or electronic format at the place of work.
	Reference documentation	To be sourced from appropriate sources and made available in printed and / or electronic format at the place of work.
	Safety data sheets	To be provided by substance suppliers and to be made available in printed format at the place of work.
	Risk assessments	To be developed by competent persons and periodically reviewed. Made available in printed and / or electronic format at the place of work.
	Emergency arrangements	To be developed by competent persons and periodically reviewed. Made available in printed and / or electronic format at the place of work. All employees to be made aware of emergency arrangements in drills, exercises and during safety inductions.
	Drills and emergency exercises	Drills to be conducted according to legal and company requirements. Debriefs to be held and minutes to be made available. Any remedial actions to be assigned and tracked.
	Remedial actions (from all sources)	Actions to be assigned and tracked and made available via hard copy and / or electronic remedial action plan.

	Information	Method
	Safety statistics[15]	Statistics to be correlated and analysed by competent persons and to be made available in printed and / or electronic format at the place of work.
	Audits, inspections and management reviews	Audits and reviews to be conducted according to legal and company requirements. Reports to be generated and to be made available. Any remedial actions to be assigned and tracked.
	Accident and incident investigations[16]	Investigations to be conducted according to legal and company requirements. To be made available in printed and / or electronic format at the place of work.
	Health surveillance reports[15]	Surveillance to be conducted according to legal and company requirements. To be made available in printed and / or electronic format at the place of work.
	Hazard and safety information[17]	Located at appropriate locations indicating existing hazards or other safety information.
Representative	Safety representative[18]	Representatives selected by employee vote according to the local requirement.
	Safety meetings	Periodic scheduled safety meetings at the place of work where agendas are developed, minutes taken and attendance recorded.

---

[15] Including leading and lagging indicators, trend analysis, etc.

[16] Procedures for the distribution of information on accident investigations and health surveillance records must take into consideration data protection and confidentiality requirements of legislation in your jurisdiction.

[17] Safety signs are a legal requirement in most jurisdictions and are used where hazards cannot be avoided or reduced by collective precautions (that protect everybody) or safer ways of doing the work.

[18] Selected according to the requirements of applicable regulations.

Information	Method
Safety committee meetings	Periodic scheduled safety committee meetings at the place of work where agendas are developed, minutes taken and attendance recorded.
Toolbox talks	Periodic toolbox talks at the place of work.

ILO-OSH 2001	3.6. Communication

See also

Q7   What is a policy statement?
Q12   How do we communicate our occupational safety and health policy statement?
Q22   How do we control safety management system documents?
Q24   How do we format a work procedure?
Q25   What are management procedures?
Q27   What is communication?
Q29   Who should receive health and safety information?
Q40   What are safety plans?
Q41   What is risk assessment?
Q51   What is a hazard register?
Q52   What is a permit-to-work system?
Q57   What is a bridging document?
Q63   How do we plan for emergencies?
Q85   Why should we analyse safety statistics?
Q90   How do we investigate accidents?
Q92   What is an audit?
Q97   What are corrective actions?

Q29 Who should receive health and safety information?

In many jurisdictions, there is a legal requirement to make certain safety management information available to all your employees and, where appropriate, to others who may be affected by your activities, such as contractors – for example, risk assessments, occupational noise assessments, etc.

However, you must consider:

- The confidentiality requirement for personal employee data that may be included in employment records, accident investigations, drug and alcohol testing results, health surveillance and medically-related issues involving employees. The minimum data protection and employment legislative requirement in your jurisdiction must be incorporated into your safety management system.

- The confidentiality requirement for corporate intellectual property, business information, contracts and other legal or financial information that may be included in specific audits, inspections and reviews.

- Reports and investigations that have been written to the draft stage but have not been finalised by a documented review and approval process. Significant problems for management can occur where statements, conclusions, remedial actions or other sensitive information that has still not been approved gets into the public domain.

ILO-OSH 2001	3.6. Communication

See also

Q15 Who is responsible for safety in an organisation?

Q28   What information needs to be communicated within a safety management system and how?

PLANNING & IMPLEMENTATION

Q30   What is an initial review?

The initial review is designed to establish a baseline of the existing occupational safety and health (OSH) arrangements in place within an organisation and to identify any gaps in the system against a chosen safety management model. Even for organisations starting from scratch, an initial review using the *ILO-OSH 2001* model can serve as a basis for developing an action plan.

Consider the following issues when undertaking an initial review of your organisation:

- Identify the regulatory, industry-specific, voluntary and senior management requirements that your safety management system (SMS) should meet (bearing in mind that, as a minimum, the system must comply with applicable laws and regulations relating to OSH in your jurisdiction).

- Identify and assess the hazards across your operations.

- Assess whether existing controls are adequate to manage risk (including the analysis of worker health surveillance, if appropriate).

- If you are starting from scratch, establish which safety management model you plan to implement, as this will provide a detailed framework against which you can assess your existing arrangements.

You need to ensure that the person(s) undertaking the review process are competent to do so and will work not only with senior management but with employees and their representatives, where appropriate.

Once the review has been completed:

- The results must be documented.

- The results can be developed into an action plan where deficiencies are prioritised, assigned timelines, the appropriate resources and a responsible person(s) to follow-up.

- The action plan can be used as a measure of continual improvement for the management of the organisation.

| *ILO-OSH 2001* | 3.7. Initial Review |

See also

Q3   How is a safety management system organised?
Q31   What should be in an initial review?
Q41   What is risk assessment?
Q99   What is continual improvement?

Q31   What should be in an initial review?

Since the initial review is essentially an audit and assessment process, you need a basic structure to follow – for example:

- Identify which occupational safety and health (OSH) laws and regulations, industry-specific best practice guides and / or voluntary codes of practice apply to your industry sector. For example, in many jurisdictions, specific OSH requirements apply to the following sectors, among others:
  - Agriculture.
  - Aviation.
  - Construction.
  - Diving and underwater activities.
  - Education.
  - Entertainment (night clubs, pubs and bars, etc).
  - Fishing.
  - Food safety (manufacture, wholesale and retail).
  - Forestry.
  - Large scale storage of hazardous substances.
  - Marine activities.
  - Mining.
  - Nuclear.
  - Offshore activities.
  - Pharmaceutical activities.
  - Quarrying.
  - Security.
- Identify the scope of your existing safety management system (SMS). If you have a formally-defined SMS, use the basic elements of the system as your initial review template. Where no formally defined system exists, use the *ILO-OSH 2001 Guidelines on Occupational Safety and Health Management Systems* structure to place your own activities and systems into a template.
- Assess identified hazards, risks and controls against your general and specific regulatory requirements to see to what extent:

- They are defined and documented within your organisation.
- They meet applicable regulatory, industry sector or voluntary standard(s).
- They meet your own specific SMS (if you have one).
- Areas have been identified where remedial work is required to fill gaps in the system.

The result of the initial review is two data-sets:

- The requirements to be met.
- The extent to which they are being met.

Next, you must consider how these two elements must now be drawn together. Success in establishing a safety management baseline from the initial review process depends on the breadth and depth of the findings, on management commitment to address issues and the time, personnel and finance resources that are made available.

Depending on the scope and complexity of your operations, it may be more practicable for the initial review to be carried out by a safety professional since there are many separate components of a SMS that needs to be looked at for the review to be of benefit to your organisation.

ILO-OSH 2001	3.7. Initial Review

See also

Q3   How is a safety management system organised?
Q30   What is an initial review?
Q41   What is risk assessment?

Q32   What is meant by planning, development and implementation?

Planning, development and implementation means the creation of a safety management system (SMS) that supports:

- As a minimum, compliance with national laws and regulations.
- The elements of the organisation's occupational safety and health (OSH) management system.
- Continual improvement in OSH performance.

The elements of your SMS that you will need to develop and populate will depend on the specific model or template that your organisation chooses to use. Although most safety management models have the same basic elements, development and implementation cannot be undertaken until the chosen model has been agreed at senior management levels, communicated within your organisation and the exact structure has been defined.

After the SMS has been developed and implemented, it cannot remain a static entity if it is to be effective in managing OSH risk; it must evolve and develop over time, through a continual improvement process, to meet your organisation's objectives as stated in policy statements and ongoing OSH objectives.

ILO-OSH 2001	3.8. System Planning, Development and Implementation

See also

Q3   How is a safety management system organised?

Q33   What safety standards should we adopt?

A number of variables will determine the standards on which to base the content of your safety management system (SMS). In order of importance, these are:

- Meeting and exceeding the minimum legal occupational safety and health (OSH) requirements applicable to the jurisdiction(s) in which your organisation operates.

- Meeting and exceeding the requirements of good practice and Codes of Practice where these are applicable to your operations.

- Meeting and exceeding the minimum industry standard requirements applicable to your business or industry sector(s).

- Meeting and exceeding the minimum safety requirements set by the Board and senior management.

Legal requirements
The legal requirement is the absolute minimum standard that needs to be defined within your SMS and all of the functions within your SMS should be assessed and developed to reflect this. For a business based in a single location, it is easy to define the minimum standard that applies to the business but, for organisations that operate in a number of different states, provinces or even different countries, it can be complicated to identify all of the requirements that might apply.

Therefore, many organisations that operate in multiple locations set the content of their SMS to meet and exceed the *highest* legal standard that applies to their operations, which in nearly all cases will satisfy and exceed the minimum requirements elsewhere. Ideally, your organisation should aspire to exceed legal requirements as, in many areas, these are only minimum standards that can be improved upon easily.

Good practice and Codes of Practice
Compliance with industry good practice and relevant Codes of Practice (CoP) (where available) is an excellent benchmark to integrate into a

SMS, as generally these offer a higher standard of guidance than the basic legal requirements.

Industry requirements

There are many industry sectors where national and international representative trade associations and organisations provide excellent guidance on minimum OSH standards and promote continual improvement in health and safety within their industry sector.

Improvements often are based upon:

- Developments among their member organisations in safety management practice and techniques.
- Published technical papers.
- Lessons learnt from incidents and accidents.
- Industry sector training.
- Development of industry equipment and operational standards.
- Tracking improvements in national and international industry sector legislation worldwide.

Senior management requirements

The standards to which an organisation operates is a reflection of the commitment to safety from the Board and / or from senior management and, therefore, it is vital that the highest safety management standards are adopted at the outset.

ILO-OSH 2001	3.8. System Planning, Development and Implementation

See also

Q3   How is a safety management system organised?

Q34   What are safety management objectives?

Safety management objectives are a useful tool for providing a measure of safety performance that can be tailored to specific operational areas, client projects or across various time periods. They should be focused towards continually improving workers' occupational safety and health (OSH) protection to achieve the best OSH performance within the organisation.

Setting performance criteria allows management to:

- Measure the safety performance of the organisation by highlighting trends.
- Focus safety activity on OSH target areas within an organisation.
- Assess OSH performance across projects carried out within an organisation.
- Assess OSH performance of contractors.
- Provide historical statistical data on safety performance.
- Benchmark performance against similar organisations or industry-wide performance standards.

Where OSH objectives are used year-on-year, whenever possible the same criteria should be used to ensure accurate long-term comparisons.

When setting objectives, it is important to assign responsibility for them. Who is responsible depends on the criteria on which the objective is based. For example, a business unit may set project-specific OSH objectives, so the project manager may be assigned the overall responsibility (and therefore accountability) to meet or improve upon the objectives set; corporate OSH objectives normally are the responsibility of senior management, who are accountable to the organisation's Board.

The nature of your organisation's business, the industry that it operates in and the criteria used within the organisation to assess performance are all factors in developing your own safety objectives. Whatever criteria are used to set these objectives, you should ensure that they are SMART –

that is: **S** – Specific; **M** – Measurable; **A** – Attainable; **R** – Realistic; and **T** – Time-bound.

ILO-OSH 2001	3.9. Objectives

See also

Q3   How is a safety management system organised?
Q15   Who is responsible for safety in an organisation?
Q35   How do we decide what occupational safety and health objectives are required?
Q36   When should we review our occupational safety and health objectives?
Q83   What is occupational safety and health performance?
Q84   Why should we measure occupational safety and health performance?
Q85   Why should we analyse safety statistics?

Q35   How do we decide what occupational safety and health objectives are required?

Areas where occupational safety and health (OSH) objectives can be set for your organisation are wide-ranging, but the following may provide some guidance on target-setting:

- Senior management participation and commitment: Participation on audit teams; visits to remote or foreign work locations; attendance at general safety or safety committee meetings; attendance at training events; fulfilling training schedule requirements.

- Standards: Adopting improved safety standards within the organisation.

- Financial: Allocation of increased financial resources for OSH budgets.

- Development of infrastructure: Implementation or roll-out of new safety equipment or systems; replacement of existing equipment with new equipment manufactured or fitted out to meet improved safety standards.

- Training: Increasing the range of training available; development of new training programmes; number of attendees fulfilling safety training requirements.

- Comparison with industry or national OSH statistics: Setting internal targets for accident and incident rates; percentage improvements on previous years' safety performance.

- Audit (internal and external): Number / percentage of completed internal and external audits carried out.

- Safety performance: Percentage of remedial actions (or action items) closed out within their specified time period; percentage reduction of remedial actions closed out beyond their due date.

Organisations that operate across the oil and gas exploration and production industry often subscribe to the OGP (International Association of Oil and Gas Producers) Health and Safety Incident Reporting System. In 2005,[19] statistical data showed that one industry safety metric called the Lost Time Injury Frequency rate (LTIF)[20] decreased from 1.09 in 2004 to 0.97 in 2005 across the industry. If your organisation subscribed to this reporting system, one of your safety objectives might be: *"To achieve an industry standard LTIF rate less than or equal to 0.97 (or the current industry average)"*.

ILO-OSH 2001	3.9. Objectives

See also

Q20 How do we develop safety training?
Q34 What are safety management objectives?
Q36 When should we review our occupational safety and health objectives?
Q59 What is management of change?
Q83 What is occupational safety and health performance?
Q84 Why should we measure occupational safety and health performance?
Q85 Why should we analyse safety statistics?
Q92 What is an audit?
Q97 What are corrective actions?

---

[19] OGP (2006). *Safety Performance Indicators – 2005 data* (Report 379).
[20] LTIF is the number of lost time injuries (fatalities + lost workday cases) per 1,000,000 hours worked.

Q36 When should we review our occupational safety and health objectives?

The review process to establish the organisation's success or otherwise in meeting occupational safety and health (OSH) objectives should be integrated into the safety management system (SMS), so that it is a scheduled management activity as part of the normal operation of the organisation. The review process depends on the time and performance criteria set for individual objectives, which in turn determine when objectives should be reviewed.

Since objectives are set as measures of OSH performance, the review process should:

- Where objectives have not been met, analyse whether:
  - Objectives were realistic and achievable when originally set (taking the SMART criteria into account).
  - Adequate resources were made available, such as finance, personnel, equipment, etc.
  - Unanticipated changes occurred during the OSH objective period, such as extreme weather conditions, civil unrest or unexpected industrial action, etc.
  - Management responsible for meeting the objectives performed adequately in their planning and implementation for the project or operations covered.
  - The SMS itself has failings that contributed towards objectives not being met, such as insufficient resources for training, lack of documented procedures, etc.
- Where objectives have been met, consider:
  - Whether targets were too easy to achieve and thus did not result in any tangible improvement over past performance results.
  - Acknowledgement (and perhaps reward) of both workers and management for their success in meeting targets.

One of the best places within your SMS to record the organisation's performance in meeting OSH objectives is in the management review document. This annual process provides a natural focal point for reviewing OSH objectives and, even where an objective is still current during a management review, it can be reviewed and commented upon as work-in-progress.

The overall benefits of reviewing objectives are that measuring performance against set criteria year-on-year identifies trends within the organisation and your own performance against industry, national and international standards, which allows management to make strategic decisions for improving safety management performance.

ILO-OSH 2001	3.9. Objectives

See also

Q3   How is a safety management system organised?
Q34   What are safety management objectives?
Q35   How do we decide what occupational safety and health objectives are required?
Q83   What is occupational safety and health performance?
Q84   Why should we measure occupational safety and health performance?
Q85   Why should we analyse safety statistics?
Q96   What is a management review?

Q37   What is planned preventative maintenance?

Equipment and systems maintenance is a process that continues to develop in a systematic way within industry. Planned preventative maintenance (PPM) (also known as fixed time maintenance (FTM)) is an evolution from the most basic and unsophisticated 'breakdown' or 'failure' maintenance approach, which replaces or repairs equipment only when it breaks.

PPM has developed as a time-based strategy, the general purpose of which is to minimise breakdowns and excessive depreciation, through the efficient scheduling of inspections to maintain or replace components before they fail, and to maximise the efficient use of maintenance staff.

The primary objectives of a documented PPM system are to:

- Reduce production downtime due to equipment breakdowns.
- Maintain equipment to relevant legislative or internally-defined safety standards.
- Increase equipment efficiency, system reliability and increased asset life expectancy.
- Maximize fuel and / or energy efficiency.
- Reduce the likelihood of unplanned 'fault' or 'corrective' maintenance.
- Reduce financial loss due to production delays.

The safety and cost implications for operating equipment maintained under a documented system are obvious and ensure that an organisation is working towards meeting its general legislative requirement to provide safe work equipment. PPM systems can be run as stand-alone systems or as part of an asset management / plant integrity programme but, in addition, they should be defined (usually as a management procedure) as a risk reduction measure within your organisation's safety management system.

ILO-OSH 2001   3.10.1. Prevention and Control Measures

See also

Q3   How is a safety management system organised?
Q25   What are management procedures?
Q38   What are the key elements of a planned preventative maintenance system?
Q39   What can be covered under a planned preventative maintenance system?

Q38 What are the key elements of a planned preventative maintenance system?

An effective and efficient planned preventative maintenance (PPM) system should include all of the following elements:

- Equipment inventory: All relevant equipment should be identified and referenced within the PPM system.

- Definition of maintenance tasks: Specific maintenance tasks required for the equipment as detailed within the manufacturer's specifications, including:
  - Inspections.
  - Adjustments.
  - Testing.
  - Calibration.
  - Rebuild.
  - Replacement.

- Frequency of internal maintenance tasks: The frequency of maintenance tasks carried out internally (including a reminder system for repetitive maintenance activates) depends on:
  - Manufacturer's recommendations.
  - Use of the component or equipment (equipment under heavy usage will be more prone to wear out).
  - Criticality of the component or equipment.
  - Legislative or best practice requirements.

- Frequency of external maintenance tasks: The frequency of maintenance tasks carried out by external competent third parties, such as the calibration, testing and certification of specialised equipment, should be defined. The providers of such services and the records generated from their external maintenance procedures (certificates, etc) should be recorded within the PPM system.

- Spares and replacement inventory: An inventory / stock of equipment spares is required to enable component replacement to take place within the defined timescales set by the PPM system.

- Tools and test equipment inventory: Specialist tools, including test and calibration equipment, may need to be tracked in the PPM system too, to ensure that they are certified or calibrated periodically by a third party.

- Review: Where the PPM process is defined within your safety management system, ensure that this process is covered by the internal audit and review system to verify periodically that the system is working as defined.

Also consider elements of an approach called 'condition-based monitoring' in your maintenance programme to identify items that cost less to replace when they fail than the cost of ongoing preventative maintenance (including maintaining spares inventory, time out of service, etc). Condition-based monitoring differs from the PPM approach in that it uses sensors, instrumentation and other monitoring and testing protocols to generate real-time data on the actual status of equipment so that decisions can be made as to the optimal time to replace or to carry out maintenance, irrespective of a defined time-line.

ILO-OSH 2001	3.10.1. Prevention and Control Measures

See also

Q37   What is planned preventative maintenance?
Q39   What can be covered under a planned preventative maintenance system?
Q92   What is an audit?

Q39   What can be covered under a planned preventative maintenance system?

The coverage of any planned preventative maintenance (PPM) system depends on the scope of the operations and associated activities of the organisation running it, but all organisations should be aware of any work equipment under their control that should be maintained, calibrated or tested due to the specific requirements of:

- Legislation that applies in your jurisdiction (such as the requirement to test lifting equipment in the UK[21]).
- Commercial insurance providers.
- Certification bodies.

The following is a typical (but not exhaustive) list of equipment that may be covered under a PPM system:

- HVAC equipment (heating, ventilation and air-conditioning).
- Electrical installations (fixed equipment such as breaker boxes, distribution panels, etc).
- Portable electrical equipment (such as power tools, office equipment, etc.).
- Pressurised equipment (such as pneumatic and hydraulic equipment, vessels, receivers and associated pipe-work, etc).
- Emergency fire and medical equipment.
- Emergency backup systems (such as generators, UPS systems, etc).
- Transport vehicles (for use on public roads).
- Transport vehicles (for use in the workplace such as forklift trucks, etc.).
- Safety systems (such as emergency stops, interlocks, guards, sensors, etc).

---

[21]   *Lifting Operations and Lifting Equipment Regulations, 1998.*

- Personal and respiratory protective equipment.

- Lifting equipment and gear (such as cranes, wires, lifting strops, etc).

- Working-at-height equipment and gear (such as scaffolds, ladders, etc).

ILO-OSH 2001	3.10.1. Prevention and Control Measures

See also

Q37 What is planned preventative maintenance?
Q38 What are the key elements of a planned preventative maintenance system?

Q40   What are safety plans?

Although there are different types of plan, a safety plan can be defined as: *"a documented plan that outlines how an organisation will implement health and safety requirements specified in law or in a contract".*

Some organisations operate in industries where there is a requirement to generate safety plans that outline how they manage their operations in a safe manner and how they comply with relevant legislation in their sector – for example, in the UK offshore sector, offshore installations are required by law to have a Safety Case[22] and, in Ireland, all businesses must have a Safety Statement.[23]

Where there is no legal requirement, a client may require safety plans from their contractors as a contractual requirement, usually based on industry best practice as recognised by the relevant industry representative organisation.[24]

Even where there is not an explicit legislative or contractual requirement, safety plans can be a useful tool for many businesses to provide concise and relevant occupational safety and health (OSH) information about projects and work locations to employees, clients and contractors. Safety plans do not have to meet any particular format or content (unless there is a legal or contractual requirement), so you have flexibility in developing your own plan specific and meaningful to your organisation.

Outside of the legal and contractual context, project-based safety plans can provide a focal point to address OSH issues in operational and production projects. Their content may include:

- Project overview.

---

[22]   Required under the UK *Offshore Installations (Safety Case) Regulations, 2005.*
[23]   Required for employers under the Irish *Safety, Health and Welfare at Work Act, 2005*; outlines how OSH issues are managed within a business.
[24]   For example, the contractual requirement for a health and safety plan to be provided by contractors in the onshore and offshore geophysical exploration industry as outlined in *OGP-HSE Guidelines for Working Together in a Contract Environment – Report No. 6.64/291*, 1999, p.9.

- Mobilisation and demob issues.
- Training and competency requirements.
- Project risk assessments and method statements.
- Timelines and schedules.
- Standards and specifications.
- Logistics.
- Journey management.

Further, for organisations with multiple work sites, location-specific safety plans can provide information on:

- Facility or plant layout.
- Safety inductions.
- Emergency equipment and system locations.
- Emergency response procedures.
- Welfare facilities.
- Waste management facilities.
- OSH focal points.

ILO-OSH 2001	3.10.1.2. Prevention and Control Measures

See also

Q19   How do we define competence?
Q20   How do we develop safety training?
Q28   What information needs to be communicated within a safety
        management system and how?
Q41   What is risk assessment?
Q63   How do we plan for emergencies?
Q73   How do we control our contractors?

Q41   What is risk assessment?

Risk assessment can be used to mean:

- A particular method or process.
- A general action that can be carried out using any one of a range of risk assessment methodologies (such as HAZID[25] and SJA[26]) .

In **QUICK WIN SAFETY MANAGEMENT**, we use the term to mean a specific process or method that follows the basic five-step process outlined below.

One of the most misunderstood elements of safety management is the requirement to carry out risk assessments in the workplace (this is a requirement across most national occupational safety and health legislation). The common perception amongst many business people is that assessing risk is a complex process; however, this is not the case. The risk assessment principle is so basic that we all make multiple risk-based decisions on a daily basis, probably without appreciating what is we are doing (if you have crossed a busy road or driven a car, then you are already familiar with the basic risk assessment process). Although there are a number of different methodologies in use (often specialised and tailored to be industry-specific or work activity-specific), for most SMEs, the basic risk assessment process outlined below is more than adequate for managing day-to-day risk.

The five steps in the process are:

- Identify the hazards.
- Decide who may be harmed and how.

---

[25] Hazard Identification: a multi-disciplinary team-based hazard analysis technique used across many industry sectors to identify and control significant hazards that can occur within a system (such as an oil pipeline, process pipework, etc) usually at the design or planning stage.

[26] Safe Job Analysis: a risk assessment technique that details specific steps in an activity, and lists the hazards and required controls at each step of the activity. Also known as Job Safety Analysis (JSA).

- Evaluate the risks and decide on precautions.
- Write down your findings.
- Review and update your assessments.

Identify the hazards
Walk around your workplace to identify hazards that could cause an accident. Ask your employees and safety representatives for their opinions, too.

Decide who may be harmed
Identify individuals or groups who could be harmed. For example, will the general public be affected? Or lone workers, young employees, contractors, foreign workers? Do you share a premises? This will help to target controls suitable for the groups identified.

Evaluate the risks and decide on controls
In general terms, employers have to do what is 'reasonably practicable' to manage risk; that is, to put in place the necessary protective and preventive measures (after identifying the hazards and assessing the risks) but not any further measures that are grossly disproportionate to the risk.

Consider the following steps (or combinations of these) when evaluating the risks in what is known as a 'hierarchy of control' but **always** start at the top:

- Eliminate the risk.
- Control the risk at source, through the use of engineering controls or organisational measures.
- Minimise the risk by the design of safe work systems, which include administrative control measures.
- Where residual risks cannot be controlled by collective measures, the employer should provide for appropriate personal protective

equipment, including clothing, at no cost, and should implement measures to ensure its use and maintenance.[27]

Involve the work force in this process as their buy-in will improve the effectiveness of agreed controls.

Write down your findings
Write down the results of your assessments. Ensure that your employees are aware of the issues raised and the relevant controls. Indicate what hazards were identified, who was affected, what controls were put in place and note that the work-force were involved in the process.

Review and update your assessments
Ensure your assessments remain current and, when changes and improvements are made in your place of work, that the risk assessments are reviewed and updated as appropriate. For example, if a new guard is fitted to a machine, update the appropriate risk assessment that it has been done, when and by whom.

ILO-OSH 2001	3.10.1.2. Prevention and Control Measures

See also

Q41   What is risk assessment?
Q42   Why is the risk assessment process so important?
Q43   What are the potential problems with risk assessment?
Q44   What does 'reasonably practicable' mean?
Q45   What is a hierarchy of control?
Q46   How should we use personal protective equipment?
Q59   What is management of change?
Q86   Why do accidents happen?

---

[27]   *ILO-OSH 2001*, p11.

Q42   Why is the risk assessment process so important?

The risk assessment process is important because:

*   Occupational safety and health (OSH) legislation requires an effective risk assessment process.
*   There is an ethical imperative.
*   There is a financial cost to a poor risk assessment methodology.

Most OSH legislation is based upon the risk assessment process, where it is the duty of the employer to ensure that workplace hazards are identified and managed. In many jurisdictions, corporations, company directors and senior management can be held personally and / or collectively liable for serious breaches or failures in implementing the requirements of OSH law.[28]

In addition to legal requirements, there is also an ethical (or moral) imperative to managing risk. Employers create risk in the workplace, such as on the shop-floor with machinery, perhaps requiring employees to drive vehicles on the public roads or working with hazardous substances. A business that does the 'right thing' when dealing with all of its stakeholders enhances its reputation in business and within the community; conversely, in the event of a significant accident or incident, the potential loss of reputation can have a significant impact on an organisation's bottom line.

If employers do not understand how to manage workplace hazards or their obligations under OSH legislation, then almost certainly there will be a financial price to be paid in the event of an accident or incident. Extensive research carried out throughout the industrialised world shows time and again that accidents can be a significant cost to a business, often resulting in fines for breaches in health and safety law, costs due to lost production, compensation payments, court fees, replacement of staff

---

[28]   For example, the UK *Corporate Manslaughter and Corporate Homicide Act, 2007.*

and increased insurance costs. It is interesting to note that, where accidents or incidents occur and companies are fined, money is always available to correct the fault, in addition to paying fines and compensation after a court case. Surely it makes more financial sense to spend a usually much lesser sum to ensure that workplace hazards are identified and managed effectively as part of the normal business processes?

To illustrate this point, the following comments are taken directly from the *Fatal Accident Investigation Report*[29] for the BP Texas City refinery explosion, which occurred on 25 March 2005, killing 15 people and injuring over 170 people, as a result of a fire and explosion on the Isomerization plant (ISOM) at the refinery:

- *"... process safety knowledge and skills within management and the workforce were generally poor. This was most evident in the area of risk awareness, where hazard / risk identification skills appeared to be generally poor throughout the supervision and crew of the ISOM unit".*

- *"The poor understanding of risk is also reflected in some of the Process Hazard Analyses ...".*

- *"There were no plans to systematically reduce safety risks in the refinery. No individual or group seems to have the accountability for driving process risk reduction across the site".*

- *"Site management did not appear to be focused on understanding and reducing the highest risks".*

- *"When risks were identified, management and the workforce appeared to tolerate a high level of risk. The investigation team observed many examples of a high level of risk being accepted within the site".*

---

[29] BP's *Fatal Accident Investigation Report, Isomerization Unit Explosion, Texas City, Texas, USA*, available at http://www.bp.com/liveassets/bp_internet/us/ bp_us_english/STAGING/local_assets/downloads/t/final_report.pdf. Extracts are from Section 5.19.1 Risk Awareness and Section 5.19.2 Risk Acceptance.

So what was the cost? In October 2009, the US Department of Labor's Occupational Safety and Health Administration (OSHA) issued a record fine of $87.43M to BP Products North America Inc. for the company's failure to correct potential hazards faced by employees.

Obviously, this is an extreme example of a major accident with multiple fatalities, but the lessons for all organisations are clear. Understanding the process of hazard identification and the implementation of controls is fundamental to manage risks effectively is fundamental to ensuring the safety, health and welfare of employees, to meeting legal obligations and to avoiding unexpected and sometimes significant costs.

ILO-OSH 2001	3.10.1. Prevention and Control Measures

See also

Q41   What is risk assessment?
Q43   What are the potential problems with risk assessment?
Q86   Why do accidents happen?

Q43   What are the potential problems with risk assessment?

The risk assessment process can bring huge benefits to your organisation but consider the following in your risk assessment options:

- Don't over-complicate things: Keep your assessment as simple as possible in order to achieve the level of risk management required for your circumstances.

- Recognise that risk assessments can be subjective: Nonetheless, most everyday hazards and their controls are already well understood and documented, so, where you need help, guidance is available, especially online.

- Don't try for perfection: Recognise that there will always be some residual risk. The aim is to manage risk and to do what is reasonably practicable. Don't try for perfection, because it doesn't exist.

- Be wary of starting your controls by issuing personal protective equipment (PPE): PPE is the control of last resort for the following reasons:
  - It only protects the individual, not the collective.
  - If it fails, you (or your staff) are exposed to the hazard.
  - You cannot guarantee correct fitting or compliance when it is required to be worn.
  - It can wear out and / or be incompatible with other PPE items.

- Keeping up-to-date: Risk assessments must be kept up-to-date and reviewed, especially if you change equipment, move premises, open a new facility or change a work process or work flow. Out-of-date assessments are of little to no value, so aim to review them annually as a minimum, unless there are significant changes in the meantime.

| *ILO-OSH 2001* | 3.10.1. Prevention and Control Measures |

See also

Q41 What is risk assessment?
Q42 Why is the risk assessment process so important?
Q44 What does 'reasonably practicable' mean?
Q46 How should we use personal protective equipment?
Q59 What is management of change?

Q44   What does 'reasonably practicable' mean?

'Reasonably practicable' defines an important principle when thinking about risk assessment and is defined in legislation as a compliance standard, although not in all jurisdictions.

All risk can be reduced beyond what would be considered a tolerable level, although there will be a cost associated with that risk reduction in terms of finance, time and probably many other resources. Achieving what is 'reasonably practicable' means finding that often difficult compromise between the cost of implementing controls and the benefits of reduced risks. It is a principle that needs to be carefully considered because, although it is a judgment call to strike the correct balance, there is much formal guidance that must be taken into consideration when looking to show that an organisation has done what is 'reasonably practicable'.

To satisfy the 'reasonably practicable' benchmark, you must ensure that, where there is formal safety guidance issued by the competent authority in your jurisdiction (such as Approved Codes of Practice (ACoP) as used in the UK and Ireland), you either follow that guidance (where it applies to your organisation's activities) or show that you have complied with the law in some other way.

Take the example of a factory that operates with old machinery. Although developments in design and technology have raised the level of safety for new machinery, it would not be 'reasonably practicable' to remove all of the old equipment and replace it with new equipment because the cost would be grossly disproportionate to the resulting reduction in risk. However, to ensure that the risk in this factory is kept as low as reasonably practicable, you might consider upgrading the old machines, systematically replacing them over time, but also consider upgrading machines in the meantime with new safety devices or other similar lower cost risk reduction measures to reduce the risk when using them.

The higher the initial risk at your place of work, the more rigorous your risk assessment should be in determining what controls are required to ensure that you have reduced the risk to as low as it reasonably practicable.

Be careful on terminology here because, depending on the jurisdiction in which your organisation operates, different terms (such as 'good practice', 'reasonably practicable', etc) may be defined in within the law in different ways (or may be not defined at all). Therefore, you need to clarify the legal terminology that applies wherever you are working and consider it in the context of your own safety management system.

ILO-OSH 2001	3.10.1. Prevention and Control Measures

See also

Q41   What is risk assessment?

Q45   What is a hierarchy of control?

The 'user manual' for driving a car is different from that of a motorbike, even though the sequence of steps is basically the same (turn on the engine, engage the clutch, select a gear, release the clutch and the brake and off you go). Likewise, the 'user manual' for managing 'working-at-height' risk is different from that of managing 'occupational noise' risk, even though the sequence of steps is the basically the same (identify the hazard, who is affected, what controls exist, what controls are required, etc).

There are a number of 'user manuals' available for recognised workplace risks and using the appropriate one as your guide to reduce risk in a specific area will provide you with a more direct and focused approach. In safety management, these 'user manuals' are called hierarchies of control and can be defined as: *"a number of controls designed to manage general or particular risks where the higher controls are considered to be the most effective, the lower steps the least effective"*.

These hierarchy of control measures are designed to reduce risk in the following way (you must always start from the top and work your way down):

- Avoid risk and, if you can't, then ...
- Prevent risk and, if you can't, then ...
- Mitigate risk using collective measures and, if you can't, then ...
- Mitigate risk using personal measures.

In the hierarchy above, we define collective measures as those that affect everyone, such as engineering controls, work procedures etc, whilst personal measures only apply to an individual, such as PPE. Unless you can eliminate a hazard and, therefore, any residual risk, the general approach often is to use a combination of hierarchy of control measures together to create an overall effective and reasonably practicable risk management approach suitable for the circumstances. So, if you cannot manage risk effectively at a particular stage of the hierarchy to a satisfactory level, move to the next level down and see what options are available there for you to use.

Note that the definition mentions both general and particular risks. The general hierarchy of control can be used as an approach to cover many health and safety risks while some activities, because of their very specific characteristics, have evolved their own specialised versions.

The most common occupational hazards that use hierarchy measures include:

- Working-at-height.
- Hazardous substances.
- Occupational noise.

Where there is a stated hierarchy of control within OSH regulations or published guidance from a competent authority for activities that you undertake, your controls for the management of your own risks should follow these.

ILO-OSH 2001	3.10.1. Prevention and Control Measures

See also

Q41   What is risk assessment?
Q44   What does 'reasonably practicable' mean?
Q46   How should we use personal protective equipment?
Q49   How can we manage occupational noise exposure?
Q50   How can we manage working-at-height?

Q46 How should we use personal protective equipment?

Personal protective equipment (PPE) tends to be popular as, generally, it is cheap and conspicuous when worn. Unfortunately, often it is used as the first resort in controlling risk because it can provide the illusion of safety to those who are unaware of its real place (at the bottom of a hierarchy of control) when managing risk. The following real life issues give more insight into the reality of using and relying on PPE.

Avian flu
A few years ago, Avian flu caused by the H5N1 virus became a worldwide concern due to the possible threat of a pandemic. Many people wore surgical face-masks to protect themselves, believing that the masks would provide some sort of protection against inhaling the airborne virus. However, surgical masks in general are ineffective against viruses such as the H5N1 virus because they are not designed for that purpose, but people still wear them in the mistaken belief that they will be protected (it's a mask, therefore it's PPE and therefore it will protect me ...).

The health-care industry-standard respirator for protection against airborne viruses is known as an N95-type respirator. But, even with a top specification mask in a test environment, these respirators are not 100% effective.[30] Although many may consider a surgical mask or N95-type respirator better than nothing in this situation, you must understand that to the extent that the PPE is not 100% effective, there is no longer any barrier between you and the hazard.

Exposure to noise at in the workplace
Many people wear hearing protection at work, either as part of their full-time work, as a visitor to a noisy worksite or as a contractor using grinders, drills and other high-noise hand-held tools. It's generally assumed that ear-plugs or ear-muffs offer protection but, for hearing protection to be as effective as possible, they must be properly fitted and

---

[30] http://euroband.com/N95_sinadequate.pdf.

be worn 100% of the time. Removing the hearing protection for as little as 10% of the time exposure (about 30 minutes in a normal working day) results in a two-thirds loss of protection![31] This shows why PPE must be treated with care.

If you must use PPE (following a risk assessment process), understand its limitations and do not be fooled into thinking that it is always a reliable and effective control, since there are too many variables involved.

ILO-OSH 2001	3.10.1. Prevention and Control Measures

See also

Q41   What is risk assessment?
Q47   Why is personal protective equipment always at the bottom of a
hierarchy of control?

---

[31]   http://www.ccohs.ca/oshanswers/prevention/ppe/ear_prot.html#_1_11.

Q47 Why is personal protective equipment always at the bottom of a hierarchy of control?

In hierarchies of control, where personal protective equipment (PPE) is mentioned, it is always at the bottom of the list – that is, it is the least effective control against a hazard.

Although PPE has a number of benefits in protecting individuals where it is appropriate – 'appropriate' meaning *"wherever there are risks to health and safety that cannot be adequately controlled in other ways"* – unfortunately, because PPE tends to be obvious, it can give the impression that a person is safer simply because they are wearing it, when at the same time other controls that are higher up the hierarchy of control (that may have a more significant impact in making a workplace actually safer) can be overlooked or even ignored.

The problems with PPE include:

- You have to comply with regulations.
- It is everywhere.
- It only protects the individual (sort of ...).
- It must all fit together.
- It is the final frontier.

You have to comply with regulations
In most jurisdictions, there are regulations that relate directly to PPE.[32] So, if you use PPE within your organisation, you will be under an obligation to comply with the requirements of the legislation regarding its use. This can involve:

- Assessing the appropriate PPE prior to its use.
- Providing prior consultation with workers and their representatives.

---

[32] For example, in the UK, within the *Protective Equipment at Work Regulations, 1992* (as amended).

- Ensuring that the PPE you use meets national and international design and manufacture standards.

- Providing it to your employees free of charge.

- Providing a choice of PPE to your employees.

- Ensuring that PPE is compatible where more than one item is used.

- Providing instruction, or even specialist training, prior to its use.

- Ensuring that it is maintained and stored properly.

- Ensuring that it is correctly used.

So, it's not just a matter of buying the stuff, you now must manage it too.

It is everywhere
PPE is relatively cheap and is available almost everywhere these days. These considerations alone can make PPE a tempting option for your organisation, if you have some workplace risk that you think requires controls, allowing higher controls to be overlooked for consideration.

It only protects the individual (sort of ...)
PPE is designed to provide protection for individuals. The degree of protection depends on a number of factors, such as its suitability for the task, fit and its general condition. The individual also has a significant influence on its effectiveness – for example, if the person does not like to wear PPE, has a poor attitude towards safety or were not given any instruction or training – there are a lot of variables at play.

It must all fit together
The effectiveness of PPE can be negated where several items of PPE are required to be worn together but are not compatible. This can be frustrating if a person has to try to fit the PPE rather than the other way around. One common example is a person who is required to wear a hard hat but also to wear hearing protection but who prefers ear-muffs to ear-plugs. Unless the hard hat is designed for the use of ear-muffs, then there is an issue.

The final frontier

Having to use PPE means that none of the other controls higher up in the hierarchy of controls list have been able to manage the risk effectively. Once the PPE becomes less than 100% effective, the user then becomes exposed to the hazard itself to a lesser or greater degree.

ILO-OSH 2001	3.10.1. Prevention and Control Measures

See also

Q45   What is a hierarchy of control?
Q46   How should we use personal protective equipment?

Q48   How can we apply hierarchy of control procedures in practice?

A hierarchy of control procedure is based on the following objectives in order of priority:

- Eliminate the risk.
- Substitute the risk.
- Isolate the hazard.
- Use engineering controls.
- Use administrative controls.
- Use personal protective equipment.

Let's see how these apply to work activities or processes using hazardous substances and to a floor that presents slip-and-trip hazards.

Control Measure	How this could apply to work activities or processes using hazardous substances	How this could apply to a floor with slip-and-trip hazards
Elimination	Can the work process or activity involving hazardous materials be eliminated? If not, can it be contracted to another company? Are new work processes available within your industry sector to eliminate the use of the hazardous substance or by-product?	Specify non-slip flooring materials at the design stage. During refurbishment, eliminate unnecessary steps or split-level floors.

Control Measure	How this could apply to work activities or processes using hazardous substances	How this could apply to a floor with slip-and-trip hazards
Substitution	Check the availability of less harmful substances to replace the existing one for use in the process or activity in the workplace (new products that may be odour-free, non-flammable, bio-degradable, have less solvent content, not harmful to the environment, etc.). Reduce the volume of hazardous substances stored to a minimum.	Replace existing floors with the latest specification non-slip materials. Use insurance-rated non-slip grade tiles and flooring materials.
Isolation	Consider whether a process and activity that uses hazardous substances can be segregated from the general workforce (including the materials themselves) to reduce smells and odours, inhalation of dust, explosion and fire risk?	Isolate or restrict access to the area to minimise the number of people using the slip hazard location.
Engineering	Use local exhaust ventilation to ensure that vapours, dust or fumes are safety extracted for work areas or storage locations. Use specialist lockers, cabinets, tanks and containers to safely store hazardous materials. Use environmental controls such as traps in drains, interceptors and spill kits.	Consider the use of non-slip mats, handrails or specialist coatings to remove or reduce slip hazards. Ensure that lighting is adequate for pedestrians.

Control Measure	How this could apply to work activities or processes using hazardous substances	How this could apply to a floor with slip-and-trip hazards
Administration	Provide a planned preventative maintenance programme to maintain equipment. Provide written procedures for employees to follow. Ensure that an observation card system is in place for defects and problems to be reported quickly. Provide safety signs appropriate to the hazard.	Highlight slip and trip hazards, use safety signs and undertake regular inspections to monitor the condition of the floors. Provide information to new employees and visitors in your workplace inductions on slip and trip hazards.
PPE	Provide gloves, aprons, respiratory personal protection, eye protection, etc as identified in your risk assessment.	Provide suitable safety footwear appropriate to the hazard.

The table shows how hierarchy of control procedures can be a useful day-today tool that can provide you with a workflow for prioritising your response when you have a risk to manage.

ILO-OSH 2001	3.10.1. Prevention and Control Measures

See also

Q45   What is a hierarchy of control?
Q49   How can we manage occupational noise exposure?
Q50   How can we manage working-at-height?

Q49 How can we manage occupational noise exposure?

The effective management of occupational noise exposure has its own hierarchy of control, based on the following objectives in order of priority:

- Reduce the noise at source.
- Reduce the noise between the source and receiver.
- Reduce the noise at the receiver.

To show how each of these elements could be used in the real world, let's see how they apply to noisy equipment.

Control Measure	How this could apply to noisy equipment
Reduce noise at source	For equipment already installed, export the process to another company and remove the machine from your production line. Develop a 'Buy Quiet' policy when purchasing new equipment. Develop maximum noise level standards for work equipment, whether leased or purchased. Use another machine with less noise output or change the work process to create less noise. Examine options for noise reduction to be fitted such as silencers, etc. Examine options for noise deadening materials to be fitted, such as in hoppers, etc. Provide planned preventative maintenance of machinery to ensure that they operate to specification. Ensure that the machine operates at its optimal performance criteria.
Reduce the noise between the source and receiver	Enclose the machine in a sound-insulating enclosure. Provide barriers or acoustic screens. Use insulating materials or other noise dampening measures on reflective surfaces, doors, etc. Relocate the machine to another more isolated area.

Control Measure	How this could apply to noisy equipment
Reduce the noise at the receiver	Provide hearing protection PPE where appropriate. Implement a job rotation scheme, minimising individual exposure times for employees in noisy areas by frequently rotating personnel. Undertake an occupational noise survey to establish noise zones. Establish clearly marked noise protection zones and use appropriate safety signs for PPE. Provide hearing tests and follow up health surveillance if appropriate.

ILO-OSH 2001	3.10.1. Prevention and Control Measures

See also

Q45   What is a hierarchy of control?
Q48   How can we apply hierarchy of control procedures in practice?
Q50   How can we manage working-at-height?

Q50   How can we manage working-at-height?

The effective management of working-at-height has its own hierarchy of control procedure, based on the following objectives in order of priority:

- Avoid working-at-height, where this is possible.
- Use work equipment or other measures to prevent falls, where you cannot avoid working-at-height.
- Where you cannot eliminate the risk of a fall, use work equipment or other measures to minimise the distance and consequences of a fall.

To show how each of these elements could be used in the real world, let's see how they apply across a wide range of common work activities.

Control Measure	How this applies across a range of common work activities
Avoid working-at-height	**Working on vehicles:** Switch from top-loading to bottom-loading tankers and trailers. **Window-cleaning:** Adopt 'reach and wash' window cleaning systems. **General:** Consider means to lower work to ground level (lighting mast gantries that can be raised and lowered down to the ground for maintenance work).

Control Measure	How this applies across a range of common work activities
Fall prevention	**General:** Consider and prioritise collective fall prevention measures over personal measures. Work from scaffold, work platforms or mobile elevated work platforms rather than ladders. Use one-way safety gates for loading supplies from forklifts onto mezzanine floors. Use purpose-designed personnel baskets and carriers when working-at-height from workplace vehicles. Provide safe working platforms when access to equipment for cleaning, maintenance or operations is required. **Construction activities:** Provide edge protection on roof work and guard rails for trenches, excavations and pits. Use crawling boards and roof ladders on roof work. **Office:** Provide suitable ladders and stepladders to prevent the use of chair, desks and similar items being used to access shelving, cupboards, etc. **Working on vehicles:** Design gantries and loading bays to prevent falls. **Working on ladders:** Instruct users in the three-points of contact system when ascending and descending ladders.
Minimise fall consequences	**Working on vehicles:** Consider fall restraint PPE, netting and other fall prevention systems. **Construction activities:** Use of safety nets and air bags. **General:** Correct use of fall protection PPE such as harnesses and shock-absorber lanyards.

ILO-OSH 2001	3.10.1. Prevention and Control Measures

See also

Q45   What is a hierarchy of control?
Q48   How can we apply hierarchy of control procedures in practice?
Q49   How can we manage occupational noise exposure?

Q51   What is a hazard register?

The term 'hazard register' has a number of different meanings depending on the reference sources used but, in **QUICK WIN SAFETY MANAGEMENT**, we define it as: *"a collection of generic risk assessment documents, which outline the significant hazards and general controls required across a range of different safety management activities that take place within an organisation's operations and for which there is a potential exposure to the organisation"*.

The documents within a hazard register should be updated to reflect the organisation's experience in hazardous activities and also to capture any regulatory requirement and best practice guidance within an industry sector.

The risk assessment process is aimed at identifying hazards and appropriate controls for specific work procedures or tasks: the hazard register provides basic outline information to be taken into account during the development of a site-specific or project-specific risk assessment. The register is useful for organisations that carry out certain repetitive tasks or activities either nationally or internationally but which need to ensure that the basic organisational minimum safety standards and industry best practice are taken into account when developing procedures at a local or project-specific level.

The following example shows a generic hazard register document concerning work equipment.

HAZARD REGISTER	
Title	Work Equipment – General Requirements
Keywords	Purchasing, maintenance, suitability, instruction, training, supervision.
Hazard description	Poor or incorrect installation. Defective and faulty equipment. Poor ergonomic design and layout. Lack of maintenance and cleaning.
What could happen?	Failure of the work equipment. Injury to persons operating, maintaining, cleaning and removing equipment.
Potential consequences	Injury to persons. Damage to equipment. Loss of production.
Controls	Purchasing policy required for work equipment to outline minimum specification requirements (such as noise and vibration levels, equipment meets minimum safety standards, etc). Installation of work equipment to be risk assessed and properly planned. Equipment to be installed and commissioned by a competent person(s). Work equipment to be covered under the organisation's documented and formal planned preventative maintenance programme. Information and instruction to be given to users prior to use (including written procedures). Adequate supervision required. Defect reporting system to be in place. Safe systems of work (such as isolation procedures and permit-to-work) to be used for the maintenance, repair, cleaning and removal of work equipment where appropriate. Appropriate personal protective equipment to be provided and worn by users when operating work equipment where appropriate.

HAZARD REGISTER	
Recovery from failure	Emergency response plan Occupational first-aiders at all work sites Contact emergency services
External References	*The Supply of Machinery (Safety) Regulations, 1992.* *Provision and Use of Work Equipment Regulations, 1998.* *HSE INDG-299 – Using Work Equipment Safely* (Rev 1 01/06). British Standard document *PD5304-2000 – Safe use of machinery*
Internal references	Management Procedure – Safe Practices (Planned Preventative Maintenance). Management Procedure – PPE. Management Procedure – Corporate Information – Responsibilities (Employees).

ILO-OSH 2001   3.10.1.2. Prevention and Control Measures

See also

Q3   How is a safety management system organised?
Q41   What is risk assessment?

Q52　What is a permit-to-work system?

The 'permit-to-work' (PTW) process is commonly understood across all industry sectors and, although there may be differences in the activities requiring a permit to work, it can be defined as: *"documented procedures that cover work activities which are considered to be hazardous enough to require additional controls above and beyond those required for normal work activities"*.

Essential elements of the PTW system are:

- Identification of the work activities considered hazardous enough to be covered under a PTW system.

- Risk assessment of these work activities and the implementation of adequate controls.

- Identification of the roles and responsibilities within the PTW system.

- Provision of personnel who are trained and competent.

- Compliance with regulations, best practice or the use of Code of Practice as the minimum standard for certain specified activities, where appropriate.

- Provision of adequate supervision.

- Plans for the management of emergency situations.

It is unlikely that an effective PTW system can work outside of the framework of a safety management system, since the PTW process relies on a number of complementary safety management functions which must be in place to support it. Examples include appropriate training and instruction, documented management and work procedures, a risk assessment methodology and process and lock-out / tag-out procedures to name but a few.

| *ILO-OSH 2001* | 3.10.1. Prevention and Control Measures |

See also

Q53   What activities require permit-to-work controls?

In any place of work where there are tasks that meet the permit-to-work (PTW) definition, these should be covered under a PTW system. In general, PTW systems are considered most appropriate for:

- Non-production work (risk-assessed tasks, such as maintenance and repair, inspection, testing, alteration, construction, dismantling, adaptation, modification, cleaning, etc).

- Routine work requiring a higher degree of control than normally would be expected of other routine tasks (risk of fire, explosion, uncontrolled release of chemicals, etc).

- Non-routine operations presenting infrequent, unusual or unexpected hazards.

There are a number of specific work tasks widespread across a range of industry sectors that have been recognised for some time as hazardous and which generally require the controls of a PTW system. These include:

- Hot work: Any work where sparks, heat or flames are generated or used (such as welding or cutting equipment, grinding with portable power tools, etc).

- Confined space entry: Any work activity that requires access to a space classed as a confined space (such as a tank, void space, deep excavation, etc).

- Working-at-height: Any work activity that requires access to work locations generally higher than two metres off the ground.[33]

- Isolation of energised equipment: Any work activity that requires the safe isolation of work equipment, which can include hydraulic, electrical, pneumatic and mechanical systems.

---

[33] The height above ground for which a PTW system may be required depends on local legal requirements in the jurisdiction where the activity takes place.

- Lifting operations: Any work involving the lifting of loads by crane or other lifting equipment.

Of course, a PTW system depends upon the industry sector – for example, for the offshore seismic industry, additional permitted work activities may include small boat operations and oilfield close approach procedures and, for the maritime industry, may include work-over-the-side procedures.

ILO-OSH 2001   3.10.1. Prevention and Control Measures

See also

Q50   How can we manage working-at-height?
Q52   What is a permit-to-work system?
Q56   What can happen when a permit-to-work system is not effective?

Q54   What is the lock-out / tag-out system and how is it used in permit-to-work activities?

Lock-out / tag-out (LOTO) is the common name for a procedure that provides for the safe isolation of energised equipment and systems. It can be defined as: *"a process of blocking the flow of energy (electrical, pneumatic, hydraulic, air, liquid, solids, etc.) to a piece of machinery, equipment or system and keeping it blocked out"*.

LOTO procedures can be developed for work on electric, pneumatic, hydraulic, gas or solid flow systems and equipment. The process ensures that the equipment being worked on does not become accidentally energised resulting in potential injury to personnel, property damage or both. Typical work tasks that may require LOTO are maintenance and repair, installation, commissioning, cleaning and decommissioning activities.

LOTO comprises two components: lock-out points / devices and tags. Isolation or LOTO points come in many forms; they can be electrical switches, mechanical valves or hydraulic valve levers. They should have the ability to be locked effectively, normally in the closed position by use of a bracket or other purpose-made device that can have a locking device attached to it. For complex work locations, these locations should numbered and be recorded in a register so that isolation points can be easily identified for specific work tasks and the appropriate lock-out or isolation device specified.

Lock-out devices used to lock-out plant and equipment can be a combination of (but not limited to):

- Fixed securing covers or guards placed over valves or other isolation points that accommodate a lockable padlock.
- Lockable scissor hasps around the isolation point.
- Cable with lockable hasps at the end.
- Trapped key systems.

In all instances and whatever device is applied, ensure that the isolation point cannot be accidentally, inadvertently or intentionally over-ridden.

The second LOTO component are tags: labels that can be written on and securely attached to isolation points to provide information. Tags come in various shapes, sizes and colours and often indicate safety or hazard messages, such as 'Out of order', 'Do not operate', 'Do not open valve', etc. Tags are NOT an effective means of isolation on their own and must never be used as such. Tags must always be applied jointly with a lock-out device.

In all instances, the following information should be entered on the tag:

- ID of the plant or equipment that is isolated.
- Reasons and remarks in relation to the isolation.
- Time and date of isolation.
- Name of designated person who applied the LOTO devices.
- Reference number to track back to a LOTO register, isolation certificate or permit.

LOTO is a very useful component that can be used in conjunction with a PTW system, where the PTW task requires equipment or systems to be isolated before the permitted work can take place. For example, if a section of pipe that would normally have a flammable product flowing through it is to be cut out of a pipe-rack using welding gear, then it would be necessary to ensure that the pipe is effectively isolated (valves locked out) at both ends to ensure that no product is accidentally released into the pipe section being removed (for example, by a valve being opened at a remote location) during the hot work operation.

| *ILO-OSH 2001* | 3.10.1. Prevention and Control Measures |

See also

Q52   What is a permit-to-work system?

Q55   How should we review our permit-to-work system?

A review of a PTW system should cover:

- PTWs, certificates and risk assessments should be retained at the site by the issuing authority for at least 30 days after completion, and then archived for a specified period to enable an effective monitoring and audit process.

- PTW monitoring checks should be undertaken by site management and supervisors to validate compliance with detailed work instructions and control measures. Information gained from permit monitoring should be used to re-inforce safe working practices on site.

- Monitoring records should be archived on site, and reviewed during periodic PTW audits.

- PTW systems should be reviewed regularly to assess their effectiveness. This review should include both leading and lagging indicators, as well as specific incidents that could relate to inadequate control of work activity.

- The PTW system should be audited regularly, by competent people, preferably external to the site or installation and who are familiar with local management system arrangements. The audit process should examine monitoring records. Non-conformance with the system should be recorded, and subsequent remedial measures tracked to ensure all issues are effectively closed out.

- Management should be notified immediately if any non-conformance is identified during routine monitoring or auditing, which cannot be immediately resolved.

| *ILO-OSH 2001* | 3.10.1. Prevention and Control Measures |

See also

Q52   What is a permit-to-work system?

Q53   What activities require permit-to-work controls?

Q56   What can happen when a permit-to-work system is not effective?

Q78   What are leading indicators?

Q79   What are lagging indicators?

Q92   What is an audit?

Q97   What are corrective actions?

Q56   What can happen when a permit-to-work system is not effective?

Properly-implemented and well-supervised permit-to-work (PTW) systems prevent accidents and, therefore, protect people, property and the environment. To prove the point, the case study below illustrates the consequences of a poorly-implemented and poorly-supervised PTW system, in one of the most tragic accidents in the UK offshore industry: the Piper Alpha disaster in the North Sea in July 1988.

A back-up propane gas condensate pump was undergoing maintenance to check its safety valve. The job could not be completed in time, so its completion was put off until the next day. The work team placed a blanking plate on the open pipe-work to seal it and left the work area. Later that evening, the primary condensate pump failed and the next work shift activated the back-up propane gas condensate pump, unaware that the maintenance work had not been completed. In the PTW system in place at the time, there was no cross-referencing to check whether the work carried out under one permit affected the work under another. When the pump was switched on, high pressure gas leaked from the hole left by the removed safety valve and caused an explosion in the processing area. This caused extensive damage to the fire-walls and resulted in oil pipe-lines rupturing. Within a short period of time, large quantities of high pressure oil were alight, with the fire spreading throughout the facility.

After about 20 minutes, the fire was out of control. The fire deluge water system had been de-activated, so it could not be used to control the fire after the initial explosion. Eventually, the high pressure gas risers (large diameter steel pipes up to 36" across operating at 2000 psi) that supplied hydrocarbon products to the rig from the other platforms in the area failed due to the heat; this resulted in another catastrophic explosion. Within 22 minutes of the initial explosion, the rig was destroyed with all of its modules falling into the sea. Of the 229 crew onboard at the time, 167 were killed.

One of the main findings of the inquiry commissioned to investigate the accident was that the permit-to-work system in use was not effectively managed and training for this safe system of work was inadequate. Flaws in the PTW system had been identified the previous year after an investigation into the death of a rigger on the platform in 1987. Perhaps if the PTW system had been reviewed and updated from the lessons learnt from the previous accident, this may have reduced the risk of another incident occurring.

There are many other examples of industrial accidents across the world that provide further evidence that poorly-implemented or poorly-supervised PTW systems can be a major contributing factor in industrial accidents.

A survey by the UK's Health & Safety Executive showed that a third of all accidents in the UK's chemical industry were maintenance-related, the largest single cause being a lack of, or deficiency in, PTW systems. Another study of small and medium-sized chemical factories found that:

- Two-thirds of companies were not checking systems adequately.
- Two-thirds of permits did not identify potential hazards adequately.
- Nearly half dealt poorly with isolation of plant, electrical equipment, etc.
- A third of permits were unclear on what personal protective clothing was needed.
- A quarter of permits did not deal adequately with formal hand-back of plant once maintenance work had finished.
- In many cases, little thought had been given to permit form design.

This is the troubling reality in many workplaces.

ILO-OSH 2001	3.10.1. Prevention and Control Measures

See also

Q52   What is a permit-to-work system?
Q53   What activities require permit-to-work controls?

Q57   What is a bridging document?

A bridging (or interface) document can be defined as: *"a documented plan that defines how diverse organisations agree on which safety management elements will be used when co-operating on a project, contract or operation"*.

Bridging documents can be a useful tool to document how organisations contracted to work together on a project or contract manage their day-to-day work activities where both have developed safety management systems (SMS). There may be a requirement within local legislation (for example, in the construction industry) to create defined interface documents for projects or contracts, but here we consider a generic document that can be tailored to individual requirements.

The objectives of a bridging document are to ensure that:

- Operations or projects are planned and conducted in line with both organisations' SMSs.
- The organisations' SMSs do not conflict with each other.
- Where aspects of both organisations' SMSs are jointly used, that the interfaces are well defined and operable (that is, the identification of which components of the individual organisations' SMSs will be used during the project, contract or operation).

When developing a bridging document, consider the following in addition to the specific requirements of the document itself:

- **Responsibilities:** Who is responsible for the creation, review and approval of the document?
- **Document revisions:** Under what circumstances will the document be reviewed.

| *ILO-OSH 2001* | 3.10.1. Prevention and Control Measures |

See also

Q28 What information needs to be communicated within a safety
management system and how?

Q58 What is in a bridging document?

Q73 How do we control our contractors?

Q76 How do we train and instruct contractors prior to work commencing?

Q58   What is in a bridging document?

In order to provide a concise and relevant bridging document, it is recommended that only those elements that need to be agreed upon are included in it, although the document should cover all of the relevant aspects that apply to the specific project or contract which may include:

- Operational interfaces:
  - Policy and objectives.
  - Reporting structure.
  - Operational and safety meetings.
  - Training and induction.
- Hazard management:
  - Risk assessment.
  - Emergency equipment and systems.
- Planning:
  - Project documentation.
  - Emergency response.
- Implementation:
  - Incident reporting.
  - Permit-to-work.
  - Personal protective equipment.
  - Planned preventative maintenance.
  - Waste management.
- Audits and inspections:
  - Operational audits.
  - Cross audits.

ILO-OSH 2001	3.10.1. Prevention and Control Measures

See also

Q57   What is in a bridging document?

Q59   What is management of change?

Management of Change (MoC) is designed to manage planned and unplanned change effectively within an organisation with the aim of minimising the risk of failure when safety-critical change occurs.

The MoC principle can be applied across all aspects of business management such as waste management and environmental issues, personnel and HR management, information technology, project management, finance, operations as well as safety management. Managing change effectively reduces workplace risk and the likelihood and severity of accidents or incidents and, therefore, can have a huge impact on profitability.

In some jurisdictions, where failings in the MoC processes have been identified as a contributory factor in serious accidents, certain high risk industries are legally required to address work procedures relating to MoC in their safety management systems.[34]

For safety management purposes, some or all of the following functions (although the list is not exhaustive) may be considered when defining where change can occur within an organisation:

- Safety-critical personnel: Managing changes to safety-critical staff at short notice due to an emergency situation.

- Organisational changes and staffing levels: Ensuring that changes to work shifts and reductions in staff levels within the organisation or to contractors do not have a negative impact.

- Operational plant and equipment: Managing changes to construction materials, design and equipment configurations.

---

[34] For example, in the US, OSHA standard number 1910.119, 'Process safety management of highly hazardous chemicals' states in section 1910.119(l)(1): *"The employer shall establish and implement written procedures to manage changes (except for 'replacements in kind') to process chemicals, technology, equipment, and procedures; and, changes to facilities that affect a covered process"*.

- Emergency systems and equipment: Ensuring the compatibility of new equipment, and that emergency systems are sufficient for when new operations or processes are introduced, etc.

- Safety systems and equipment: Managing changes such as allowing operational equipment to run with safety systems disabled or removed.

- Work procedures, drawings and related documentation: Managing changes in drawings or documentation that relate to equipment or work activities.

- Work processes: Managing changes to instrumentation, controls, computer systems and software systems.

- Project or construction management: Managing changes to specifications and work schedules.

- Legal Acts, regulations and legal guidance: Managing changes which can be the result of updates and revisions to existing legislation and / or the issuance of new legislation.

It is critical that temporary change within an organisation is covered under MoC procedures. Temporary changes (involving equipment, processes, personnel, etc) can be a contributory factor in serious accidents as there can be a tendency for them to become permanent if not properly tracked, monitored and managed.

ILO-OSH 2001	3.10.2. Management of Change

See also

Q60   What is an 'in-kind' change?
Q61   What should we consider under management of change?
Q62   How should we handle management of change?

Q60   What is an 'in-kind' change?

'In-kind' means where a new or replacement item is the same as the old one, otherwise it is 'not-in-kind' and the MoC process must be implemented before the change can be made.

Although this type of change seems to be self-evident, the following issues need to be considered:

- Standards and specifications: Replacement items must be to the same specification and standard as the original to be considered as an 'in-kind' replacement. A replacement that meets an alternative specification or standard, even if the specification is considered to be equivalent or improved, needs to be evaluated under a MoC process.

- Routine replacement: For example, if there is a regular but unusual failure within the system that has initiated the requirement for a new item, then putting in another identical unit will not address the root cause for the regular failure.

- Change to the operation: An 'in-kind' replacement should only be put into a process or system that has not been modified or changed and where all of the relevant operating parameters are the same.

ILO-OSH 2001	3.10.2. Management of Change

See also

Q59   What is management of change?

Q61   What should we consider under management of change?

The following simple criteria should be considered when defining the management of change (MoC) process within your organisation:

- Identify and assess: Look closely at the range of work activities within your organisation and identify and risk assess those areas where a planned or unplanned change could present a significant risk to your personnel or property.

- Document and define: The procedures that outline how to undertake and manage change must be written down, communicated to all relevant persons within your organisation, and be reviewed and updated periodically.

- Record: Where temporary or permanent changes have been undertaken within the MoC procedures, ensure that all records generated (such as MoC checklists, forms or change requests, etc.) are stored within the SMS for a defined period.

- Review: Where the MoC process is defined within the SMS, organisations should ensure that this process is covered under the internal audit and review system to verify periodically that the system is working as defined.

ILO-OSH 2001	3.10.2. Management of Change

See also

Q41   What is risk assessment?
Q59   What is management of change?
Q62   How should we handle management of change?

Q62   How should we handle management of change?

Although it may appear an abstract concept, management of change (MoC) has a significant role to play in reducing risks, especially in high risk industry sectors. A lack of, or ineffective, change procedures were noted as contributing factors in a number of fatal accidents, including:

Accident	What happened?	Contributory factors relating to MoC
Chemical plant fire and explosion – Flixborough, UK , June 1974	A cyclohexane plant was discovered to have a problem with one of its six reactors. Once the plant was shut down to identify the problem, it was decided to remove reactor #5 and to connect reactor #4 and #6 together. The 20" pipeline bypass system used ruptured, which resulted in a massive escape of cyclohexane. This formed a flammable vapour mixture, which found an ignition source and ignited. The explosion and subsequent fires resulted in 28 fatalities and 36 injured at the site.	A change to the reactor plant was undertaken without a full assessment of the potential consequences including: Only limited calculations were undertaken on the integrity of the bypass line. No calculations were undertaken for the dog-legged shaped line or for the bellows. No drawing of the proposed modification was produced.[35]

---

[35] http://www.hse.gov.uk/comah/sragtech/caseflixboroug74.htm.

Accident	What happened?	Contributory factors relating to MoC
Esso Longford Plant – Australia, September 1998	The warm oil supply to a heat exchanger went offline for a period of time, resulting in a drop in temperature (estimated to be -48°C) of the heat exchanger unit to well below its normal operating temperature (approximately 100°C). When the warm oil supply was resumed, the large temperature differential caused a brittle fracture in the vessel that resulted in the release of a massive hydrocarbon vapour cloud. This vapour cloud subsequently ignited and resulted in a number of fires and adjacent vessel ruptures. The explosion and subsequent fires resulted in two fatalities and eight people injured at the site.	There were significant changes in operating processes, staffing and procedures at the Esso plant without thorough risk assessments being carried out for the changes.[36]

ILO-OSH 2001	3.10.2. Management of Change

See also

Q59   What is management of change?

---

[36]   http://www.fabig.com/Accidents/Longford.htm.

Q63   How do we plan for emergencies?

Emergency planning is an essential element within safety management. Arrangements put in place, as a minimum, must meet the national legal requirements required of your organisation and should be developed to meet industry best practice. The simplicity or complexity of your emergency planning is a direct function of the size and nature of your operations and to what extent (if any) they are covered under additional national emergency response legislation, require co-ordination with external emergency resources (fire brigade, police, hospitals, etc) and with the local authorities.

When developing emergency plans, whether you have a major hazard site or have a low risk environment, you should consider your plans in the context of this statement: *"The emergency plan should address the response required during every phase of the emergency, both the immediate needs and the longer term recovery. The first few hours after the accident occurs is the 'critical' phase of an accident response, when key decisions, which will greatly affect the success of any mitigation measures, must be made under considerable pressure and within a short period of time. Therefore, a detailed understanding of the likely sequence of events and appropriate countermeasures will greatly benefit anyone who may reasonably be expected to have a role to play"*.[37]

Emergency response planning can be broken down into three elements:

- Identification of potential hazards.
- Prevention and control measures to be used.
- Mitigations to minimise the effect of an emergency.

Identification of hazards

The risk assessment process is always the starting position for emergency response planning, which must be appropriate to the size and nature of

---

[37] HSE Books (2003, reprint). *Emergency Planning for Major Accidents – COMAH 1999*, HSG191, p.5.

your operations. The assessment process should identify the potential for accidents and emergency situations.

Preventative measures

Preventative measures that can be put in place to prevent an emergency situation developing include:

- Physical measures: Ensuring that work equipment purchased is manufactured to the relevant safety standards, combustible materials stored in a safe manner and location, sources of possible ignition identified, etc.

- **Organisational measures:** Vocational training and instruction, work and maintenance procedures, planned preventative maintenance programmes, regular workplace audits and inspections, etc.

Minimising the effect of an emergency

Emergency incidents can occur in any workplace, so the measures implemented to mitigate (minimise) the effects of an emergency situation can include:

- **Hardware measures:** Fire extinguishers, hoses, emergency exits, alarms, etc.

- **Organisational measures:** Emergency procedures, communications, emergency training and instruction, drills, emergency equipment inspections and testing, adequate emergency safety signage, designated muster-stations, etc.

For organisations that may not have the required competency internally, external competent resources may be needed to assist in the development of a compliant emergency response plan. Once such a plan has been developed, it can be integrated into the organisation's safety management system, where it should be reviewed and updated as appropriate.

See also

Q3   How is a safety management system organised?
Q20   How do we develop safety training?
Q37   What is planned preventative maintenance?
Q41   What is risk assessment?
Q92   What is an audit?

Q64   How do we plan for fire safety?

Regardless of your industry sector, probably the most common emergency that can occur in your place of work is that of fire. Before considering your own fire safety options, ensure that:

Before considering fire safety management principles, ensure that:

- You identify (in your initial review process) whether any additional laws, regulations or Approved Codes of Practice relating to fire safety or emergency planning apply to your operations.[38]

- You meet any requirement to hold a valid fire safety certificate or other documentation that addresses fire safety at your (or your landlord's) premises where appropriate and that any restrictions or conditions within these documents are met.

- Fire safety is reviewed under a documented risk assessment process by a competent person.

- Fire safety risk assessments are actioned and periodically reviewed in your workplace.

The effective management of fire safety in the workplace can be broken down into five separate elements, which you should risk assess separately. The five elements are:

- Prevent fires from occurring.

- Provide means of detection and warning.

- Provide means of fire-fighting.

- Provide a safe means of escape from the premises.

- Provide emergency plans and fire safety training.

---

[38] Examples where specific requirements may apply for additional fire safety controls in businesses include nursing homes, educational premises, hotels and hostels, theatres and cinemas, venues for concerts (open air and indoor), high occupancy office blocks, business that have the potential for explosive atmospheres (such as petrol stations), etc.

You must bear in mind that all of the five elements noted above must be compatible with each other and that all of these are examined in a holistic manner, otherwise additional hazards may be introduced.

ILO-OSH 2001	3.10.3. Emergency Response

See also

Q41   What is risk assessment?
Q63   How do we plan for emergencies?
Q65   How do we prevent fires from occurring?
Q66   How do we detect fires and raise the alarm?
Q67   How do we handle fire-fighting?
Q68   What types of fire-fighting equipment are available?
Q69   What means of escape from fire should we plan?
Q70   What do we need to do for fire emergency plans and training?

Q65   How do we prevent fires from occurring?

Fire is made up of three components: oxygen, an ignition source and fuel, so the removal of any one of these can prevent a fire from starting. In practical terms, the focus must be on managing the ignition and fuel sources effectively.

The identification of ignition sources and potential fuels for fires needs to be undertaken in a planned and systematic manner, often with the use of drawings and general arrangement plans of your workplace.

Ignition sources, fuels and some possible controls within a workplace could include (but are not restricted to):

Ignition sources	Controls
Deliberate fires set by arsonists	Assess security arrangements including: Controlled access onto the premises. Use of CCTV and appropriate warning signage. Condition and suitability of all perimeter fencing. Thorough assessment and vetting of personnel when hired. Identify possible locations where fires can be easily started (such as stacked cardboard or other combustibles stored externally).
Sparks and flames from hot work activities such as welding and cutting	Hot work to be carried out in hot work zones where possible. Hot work to be carried out under a permit-to-work system.
Smoking	'No smoking' policy to be drafted and communicated. Designated smoking areas to be made available. Naked flames (including matches and lighters) to be prohibited in areas where a fire risk has been identified. Appropriate hazard and safety signage to be used.

Ignition sources	Controls
Sparks and fires for defective electrical installations (both fixed and portable)	Ensure that all fixed and portable electrical installations are identified and covered under a planned preventative maintenance (PPM) programme.
Faults with work equipment due to lack of maintenance, poor condition or incorrect use	Ensure that all work equipment is identified and covered under a PPM programme. Provide machine-specific safety inductions for all persons required to operate them.
Contractor or third party activities (where fire safety awareness levels may be lower)	Assess contractor competency prior to signing contracts. Provide adequate safety inductions including fire precautions and arrangements that apply to all contractors working at your workplace. Ensure adequate supervision of contractors on site.

Fuel sources	Controls
Combustible materials stored in vulnerable locations such as wooden pallets stacked against buildings or waste cardboard stored in large quantities internally	Provide secure storage of combustible materials. Do not allow large quantities of combustible materials to accumulate. Do not store combustible materials close to buildings, fuels tanks or other vulnerable locations.

Fuel sources	Controls
Oils, paints and other hazardous substances incorrectly stored, flammable vapours accumulating in closed or poorly ventilated spaces	Review requirement for hazardous substances in your workplace and assess whether less hazardous substances can replace existing ones. Provide secure and dedicated storage for hazardous materials. Ensure ventilation is adequate wherever vapours and gases can accumulate. Ensure good housekeeping routines are in place. Ensure that excessive amounts of combustible materials are not kept on site. Store materials that present a potential fire hazard (such as compressed gas bottles and oxidising substances) according to manufacturers' recommendations and industry best practice. Ensure suitable and sufficient safety signage is posted clearly at all locations where fuel sources are stored. Ensure the company safety induction covers high-risk fire locations and the actions that employees are required to take in an emergency. Ensure that suitable and sufficient fire fighting and fire detection systems are in place, appropriate to the risk present.
Flammable materials used for curtains, furniture and fittings	Replace with fire-retardant materials or treat with fire-retardant coatings (where appropriate).

Use a risk assessment form for documenting fire hazards in your workplace. In that way, once the fire hazards have been identified, the same form can be used to document the appropriate controls and responsibilities too.

ILO-OSH 2001 — 3.10.3. Emergency Response

See also

Q37   What is planned preventative maintenance?
Q41   What is risk assessment?
Q52   What is a permit-to-work system?
Q64   How do we plan for fire safety?
Q66   How do we detect fires and raise the alarm?
Q67   How do we handle fire-fighting?
Q68   What types of fire-fighting equipment are available?
Q69   What means of escape from fire should we plan?
Q70   What do we need to do for fire emergency plans and training?

Q66 How do we detect fires and raise the alarm?

Once the risk assessment process has been carried out and potential ignition and fuel sources have been identified, appropriate measures for detecting and warning of fires need to be assessed. Ignition and fuel sources play a significant part in determining what fire detection and fire-fighting measures are required, so it is important that a person with the required fire safety knowledge and experience is involved in the assessment process for suitable detection and warning controls.

For detection and warning controls required, bear in mind that:

- There may be minimum legal requirements for fire safety outlined on your fire certificate (where appropriate) or stated in the regulations within your jurisdiction that could apply to your business activities or to your business location.

- Arrangements for the detection and warning systems in your workplace should be documented in your organisation's safety statement or safety policy document (appropriate to your jurisdiction).

- The fire safety risk assessment that addresses detection and warning controls should be specific to your workplace and periodically reviewed.

- Detection and warning systems should be covered under a planned preventative maintenance system, with records maintained in a Fire Register.

- Detection and warning systems should be manufactured, installed and maintained to the relevant national standard.[39]

---

[39] Examples of standards for fire safety include the UK standard BS 5839-1:2002+A2:2008, *Fire Detection and Fire Alarm Systems for Buildings. Code of Practice for System Design, Installation, Commissioning and Maintenance* and the Irish standard IS 3218:2009, *Fire Detection and Fire Alarm Systems for Buildings - System Design, Installation, Servicing and Maintenance*.

- Where there is any doubt as to fire safety issues or requirements that apply at your place of work, contact your local fire authority for advice and guidance.

What factors determine the appropriate system for detecting fires and raising the alarm for a fire in a building? Consider:

- The size of the building: Is the building so large (such as a large warehouse) that an automatic fixed alarm system is the only practical way to notify all personnel of a fire or is it so small (such as a small retail shop) that verbal warnings would be sufficient?

- Unoccupied areas: Are there areas that tend to be unoccupied under normal working conditions and where a fire could remain undetected for long periods of time?

- Hearing / seeing the alarm: Can the alarm (by whatever means it is raised) be heard in all locations or are additional alarms such as flashing warning lights required (for example, in locations where there are high noise areas such as machinery spaces, etc.)?

- Purpose of the building: The detection and warning controls should be appropriate to the use of the building, which may also be outlined within legislation. For example, buildings used for accommodation (such as a hotel or nursing home) or have been designated as an environment with the potential for explosive atmospheres may have a specified minimum requirement for fire safety detection and warning systems which must be met.

In most circumstances, the equipment used to detect fires and to provide warnings in your place of work should be defined within your fire safety risk assessment. For fire detection systems, these consist of detectors (depending on the appropriate type for your specific needs) and alarms systems, either manual or automatic. If there is any doubt as to the correct type, installation method (battery or mains powered), or number and location of these types of units in your place of work, consult a competent person or authority.

- Detectors: Three main detectors types are available for the majority of work places, and each have been designed for specific uses:
  - Heat: These are designed to detect a rise in ambient temperatures and normally are set to go off above a pre-determined temperature threshold. The two main types are rate-of-rise (RoR) and fixed detectors. RoR units detect a raise in temperature from a preset baseline over a time period (for example, $12^{\circ}C$ per minute), whereas fixed units raise the alarm when a preset temperature threshold has been reached.
  - Smoke: These detectors raise the alarm where there is a presence of smoke particles in the air above a certain threshold (sensitivity of smoke detectors are measured in units called obscuration levels). The two main types are optical (or photo-electric) or ionising, but combination units are also available.
  - Flame: This type of detector is used to detect the presence of flame at a location by either the use of UV[40] and/or IR[41] technology.
- Alarms:
  - Automatic: There is a wide variety of automatic fire alarm systems, the complexity and cost of which depends on the fire risk and building size. 'Automatic' means that a warning is generated from a sensor or detector that activates an alarm without any human intervention.
  - Manual: Equipment that is activated manually, such a manual call point, gongs, buzzers or other similar alarm equipment.

In addition to fire detectors, gas detectors also can be incorporated into a fire safety system where there is a risk of unsafe (explosive) levels of gas building up within a space or work area.

Once a fire has been detected, the emergency procedures that have been put into place should address the issue of who will contact the Fire

---

[40] Ultra-violet detectors work by detecting UV light (light in the frequency range between approx. 400nm to 150nm) emitted from fires.

[41] Infra-red detectors work by detecting infra-red emissions from a hydrocarbon fire source.

Brigade or other emergency resources that are available to you when a fire alarm is activated.

ILO-OSH 2001	3.10.3. Emergency Response

See also

Q41 What is risk assessment?
Q64 How do we plan for fire safety?
Q65 How do we prevent fires from occurring?
Q67 How do we handle fire-fighting?
Q68 What types of fire-fighting equipment are available?
Q69 What means of escape from fire should we plan?

Q67   How do we handle fire-fighting?

Your fire safety risk assessment will have identified the ignition and fuel sources, the appropriate detection and warning systems, and also the fire-fighting techniques that are the best fit in your place of work. Of course, fire-fighting methods must be a match for the characteristics of the fire, since incompatibilities between these can cause significant additional hazards (a common example of the incompatibility of fuel source and fire-fighting medium is the well-known household example of using water to douse a hot oil fire).[42]

For fire-fighting equipment, bear in mind that:

- There may be minimum legal requirements for fire-fighting equipment outlined on your fire certificate (where appropriate) or stated in the regulations within your jurisdiction that could apply to your business activities or to your business location.

- Arrangements for fire-fighting equipment should be documented in your organisation's safety statement or safety policy document (appropriate to your jurisdiction).

- Your staff will need training appropriate to the fire-fighting equipment available and to their duties (such as fire warden, fire team member, etc.).

- The fire safety risk assessment that addresses fire-fighting equipment should be specific to your workplace and reviewed periodically.

- Fire-fighting equipment should be covered under a planned preventative maintenance system with records maintained in a Fire Register.

---

[42] The rapid, almost instantaneous heating of water on contact with the hot oil (which is well above the boiling point of water) causes an explosion of steam and oil, which then becomes a significant fire hazard in itself. The use of a fire blanket to smother a hot oil fire is the correct method to use, once the fuel source has been switched off.

- Appropriate safety signs should be used to clearly mark the location and type of fire-fighting equipment.

- Fire-fighting equipment (where appropriate) should be manufactured, installed and maintained to the relevant national standard.

- Where there is any doubt as to fire safety issues or requirements that apply at your place of work, contact your local fire authority for advice and guidance.

| *ILO-OSH 2001* | 3.10.3. Emergency Response |

See also

Q68   What types of fire-fighting equipment are available?

Fire-fighting equipment (FFE) can be split into two categories:

- Fixed systems: Fixed installations are permanent systems that are usually built into an area to provide fire-fighting cover for a specific fire risk and include:
    - Sprinkler systems (water and / or foam).
    - Fixed fire hoses.
    - Deluge systems.
    - Gas (such as CO_2) systems.

- Portable equipment: Equipment that can be moved around from place to place and usually can be handled by a person on their own. This type of equipment includes:
    - Portable fire extinguishers (PFEs).
    - Fire blankets.

ILO-OSH 2001	3.10.3. Emergency Response

See also

Q64   How do we plan for fire safety?
Q67   How do we handle fire-fighting?

Q69   What means of escape from fire should we plan?

At all times, the first priority in a fire situation is to prevent loss of life. It is not always possible to safely extinguish a fire, even if the alarm has been raised in time and the appropriate fire-fighting equipment is available. In this case, the timely and organised evacuation of personnel to a designated place of safety is the priority. In all cases, your fire safety risk assessment must address means of escape to a place of safety.

For a safe means of escape, bear in mind that:

- There may be minimum legal requirements for means of escape outlined on your fire certificate (where appropriate) or stated in the regulations within your jurisdiction that could apply to your business activities or to your business location.

- Arrangements for a means of escape should be documented in your organisation's safety management system.

- Your staff will need training (such as drills) and familiarisation with the means of escape within your building or facility.

- The fire safety risk assessment that addresses means of escape should be specific to your workplace and reviewed periodically.

- Equipment assisting in providing a means of escape (such as emergency lighting, smoke hoods, etc.) should be covered under a planned preventative maintenance and periodic inspection system, with records maintained in a Fire Register.

- Appropriate safety signs should be used to clearly mark routes for a safe means of escape and places of safety or refuge.

- Emergency exit equipment and systems (where appropriate) should be manufactured, installed and maintained to the relevant national standard.

- Where there is any doubt as to fire safety issues or requirements that apply at your place of work, contact your local fire authority for advice and guidance.

You also must consider the following elements relating to a means of escape when carrying out a fire risk assessment:

- Primary and secondary routes: The primary route of escape may not always be available in an emergency situation, so alternative or secondary routes need to be defined, documented and correctly signed.

- Unacceptable means of escape: Identify locations in your workplace that should not be used as an escape route in a fire emergency. These could include lifts (with certain exceptions), ladders, wall and floor hatches and self-rescue devices.

- Signs and notices: The format, standard or type of safety sign will be defined within your own jurisdiction but probably will be of a pictogram type. For means of escape, signs should be used for:
  - Emergency escape routes.
  - Fire doors.
  - Indication of emergency exits and how to open them.
  - Keep clear signs (for example, where exit routes open to outside spaces such as car parks, etc.)
  - Assembly or muster locations.

- Consideration for disabled persons: Consider the needs of disabled persons (both employees and members of the public) in your fire safety risk assessment, including provision for persons with impaired hearing, impaired vision and for wheelchair-users.

- Maintenance of escape routes, systems and equipment: All elements associated with escape from a facility should be inspected periodically, formally maintained and recorded in a Fire Register. Inspections should ensure that:
  - Escape routes are free from obstructions, slips, trip and falls, etc.
  - Fire and safety doors are in good condition, free to open and close, etc.
  - Safety signs are legible, in good condition and that none are missing.
  - External exits can be opened easily, are not blocked inside or out and that they are clearly signed.

- Emergency lighting and exit signs are operational and are tested.

- Occupancy: The number of people and their purpose within a building will influence how you organise your means of escape. You must comply with the legal requirements for occupancy limits and ensure that your risk assessment reflects controls that are suitable for your maximum occupancy levels.

- Drills and familiarisation: Fire drills are a legal requirement to ensure that your workforce understand the procedures to follow when escaping from a facility in an emergency. Vary drill scenarios, ensure that secondary routes also are used and that your workforce are familiar with the alarm tones. Circumstances vary, but it should be realistic to evacuate a building within two or three minutes (within one minute from a high risk area). Results of drills and exercises should be documented in your Fire Register and drill debriefs also should be conducted and recorded.

ILO-OSH 2001	3.10.3. Emergency Response

See also

Q20	How do we develop safety training?
Q37	What is planned preventative maintenance?
Q41	What is risk assessment?
Q64	How do we plan for fire safety?
Q65	How do we prevent fires from occurring?
Q66	How do we detect fires and raise the alarm?
Q67	How do we handle fire-fighting?

Q70   What do we need to do for fire emergency plans and training?

Emergency plans are the 'how to' in the event of a fire, drawing together all of the disparate elements (detection, raising the alarm, fire-fighting and means of escape) into a cohesive and understandable format.

Assuming all the relevant fire safety controls are in place, you need to communicate this information not only to your workforce, but also to visitors, members of the public, contractors and other third parties who may visit your site (not all of whom may be familiar with your arrangements for emergency situations). This communication can be done in a Fire Safety Plan as part of your safety management system (in the 'Hazard Prevention - Emergency prevention, preparedness and response' section of your safety management system (SMS) if you adopt the *ILO-OSH 2001* model).

For emergency plans, bear in mind that:

- There may be minimum legal requirements for emergency plans outlined on your fire certificate (where appropriate) or stated in the regulations within your jurisdiction that could apply to your business activities or to your business location.

- Arrangements for emergency plans should be documented in your organisation's SMS.

- Your staff will need training (such as drills) and familiarisation with the emergency plans within your building or facility.

- The fire safety risk assessment that addresses emergency plans should be specific to your workplace and reviewed periodically.

- Where there is any doubt as to fire safety issues or requirements that apply at your place of work, contact your local fire authority for advice and guidance.

Many variables determine your particular fire safety arrangements and emergency plans and so you will need to consider:

- Accuracy: The plan accurately reflects existing arrangements put in place after a thorough risk assessment process and will be reviewed periodically.

- Content: The use of written instructions and flow diagrams located at strategic locations in your workplace to show succinct and concise information on what to do in a fire emergency. Ensure that instructions are short, unambiguous and easily understood by everyone who should be aware of them.

- Training: Outline what information, instruction and training is required for your workforce to ensure that the plans are understood.

- Availability: Where are emergency instructions to be made available in your place of work? You will need to identify key locations where specific emergency information is made available, such as printed copies on a safety notice board, in the canteen or in the reception area. You also will need to ensure that all third parties are briefed on your fire safety arrangements (usually during a safety induction) prior to them coming onto your site.

Consider the following factors when compiling your fire safety plan:

- What to do if you discover a fire.
- What to do if you hear the alarm.
- Contacting the Fire Brigade or other emergency response resources.
- Specific actions to be taken by persons assigned emergency duties (such as fire wardens, fire team members or escorts for visitors or contractors whilst on site, etc.).
- Identification of primary and secondary escape routes.
- Location of alarm points and fire-fighting equipment.
- Location of assembly points and muster locations.
- Procedures for dealing with specific scenarios, such as:
  - Managing the evacuation of disabled persons.
  - Managing the evacuation of third parties on the premises.

- Arrangements for taking a roll-call or head-count at the assembly point.
- Arrangements for undertaking shut-downs or other emergency procedures for high risk areas, work equipment or work activities in the event of a fire.

ILO-OSH 2001	3.10.3. Emergency Response

See also

Q71   What is procurement?

Procurement can be defined as: *"the sourcing and provision of goods and services to the organisation that meet predefined standards and requirements as outlined in national regulations and which comply with the organisation's own minimum safety management specifications"*.

The supply of many goods and services can have a significant impact on the safety of personnel within an organisation. It is for this reason that there are minimum legal requirements in many jurisdictions (for example, within the EU) that outline minimum standards for work and safety equipment and for due diligence when sourcing services.

So, within your own organisation's safety management system, arrangements and procedures specifically for the procurement of goods and services should be established and maintained to ensure that:

- The organisation's own health and safety requirements are identified, evaluated and incorporated into procurement specifications, prior to the procurement of goods and services.
- National laws and regulations are identified and incorporated into procurement specifications, prior to the procurement of goods and services.
- Arrangements are made to achieve / ensure conformance to the requirements, prior to the use of the goods.

| *ILO-OSH 2001* | 3.10.4. Procurement |

See also

Q72   What needs to be covered under procurement?

Q72   What needs to be covered under procurement?

The breadth and scope of your organisation's procurement activities depend on:

- The size of your organisation and the scope of operations.
- The equipment, systems and substances that are used by or supplied to your organisation.
- The legal health and safety requirements in the jurisdiction(s) where your organisation operates.
- The use of outsourced technical and professional services.
- The use of leased or hired equipment.

Procurement can cover the sourcing, supply and purchase of a huge range of products and services, many of which can have a significant impact on health and safety. Therefore, procedures for procurement must identify the regulatory requirements and the minimum standards that apply during this process and embed these into your system.

Although not exhaustive, the following list highlights examples of a range of issues with associated health and safety implications, some of which you may need to consider within your procurement procedures:

- Standards and specification of new and used work equipment.
- Supply of hazardous substances, including safety data sheets.
- Noise and vibration emissions from tools and equipment.
- Ergonomic considerations including display screen equipment.
- Personal and respiratory protective equipment standards and specifications.[43, 44]
- Minimum competency standards of contractors and third parties.

---

[43] European Standard, as issued by the European Committee for Standardisation.
[44] The American National Standards Institute is the US-based standards and conformity assessment system.

- Emission standards and specifications for road and workplace vehicles.

- Minimum safety requirements for road vehicles.

- Minimum standards and specifications for national and international travel (accommodation, carriers, etc).

Procurement also ties in with other elements of safety management, notably management of change (MoC). It is vital that procurement procedures are integrated into the MoC process so that equipment purchased or supplied with a changed or modified specification does not introduce a safety hazard to the end-user.

| *ILO-OSH 2001* | 3.10.4. Procurement |

See also

Q71   What is procurement?

Q73   How do we control our contractors?

Contractors often play a significant role in many organisations' operations by providing a level of expertise and knowledge not available internally. It is important to manage contractors effectively to ensure that:

- Contractors are competent to carry out contract requirements.
- Work is conducted within an effective safety management framework to eliminate or to manage risk effectively.

The two key roles are that of client and contractor. A client can be defined as *"any employer in the public or private sector who uses contractors"*[45] and a contractor as *"anyone brought in by a client to work at the client's premises who is not an employee of the client"*.

In some jurisdictions and areas of activity, there may be prescriptive responsibilities defined in legislation for the organisation's role as a client and also for the contractor – for example, in construction regulations. Where this is so, these responsibilities should be outlined in a management procedure.

The effective management of contractors involves:

- Contract of work.
- Evaluating and selecting contractors.
- Planning of contractor work prior to commencement.
- Training (induction / familiarisation / orientation) prior to work commencing.
- Management and supervision of contractors.

ILO-OSH 2001	3.10.5.1. / 3.10.5.2 Contracting

---

[45]   HSE Books (2003). *Use of Contractors: A Joint Responsibility* – INDG368 11/03.

See also

Q19 How do we define competence?
Q53 What activities require permit-to-work controls?
Q59 What is management of change?

Q74   How do we evaluate and select contractors?

The evaluation and selection process often depends on the complexity and risks of the work involved. Therefore, as a client, you need to develop an evaluation process suitable for your workplace and for the work required of your contractors.

Basic information that can be used to assess contractor competency includes:

- For less complex work environments:
    - Copies of general or specialist safety training certificates for contractor's personnel.
    - Copies of vocational training certificates for contractor's personnel.
    - Copies of medical training certificates for contractor's personnel (where appropriate)
    - Testimonials from previous clients.
    - Up-to-date copy of contractor's (and sub-contractors', if appropriate) safety statement documents.
    - Copy of contractor's public and employer liability insurance certificates.
    - Proof of valid membership of professional representative organisations (where appropriate).
- For more complex work environments:
    - Pre-selection audit that qualifies a contractor to bid for work with the client.
    - Comprehensive work tender questionnaire (including health and safety information, such a competency and training, risk assessment procedures and processes, safety management system information, health and safety statistics, accident and incident information, etc).

When using the services of a primary contractor, who then may use sub-contractors under same the contract of work, ensure that the issue of sub-contractor health and safety minimum requirements also are addressed in any contract.

ILO-OSH 2001   3.10.5.1. / 3.10.5.2 Contracting

See also

Q73   How do we control our contractors?
Q75   How do we plan contractor work prior to commencement?
Q76   How do we train and instruct contractors prior to work commencing?
Q77   How do we manage and supervise contractors?

Q75   How do we plan contractor work prior to commencement?

Regardless of the size and complexity of the proposed work, it is vital that the risk assessment process is used fully prior to any work commencing and that the scope of work is detailed enough so that all of the potential hazards are identified, assessed and understood. For example:

- For less complex work environments: Risk assessments can be carried out prior to work commencing. These should reflect the local conditions that can be seen at a work site and need to be reviewed and agreed by the client representative.

- For more complex work environments: Project plans can be developed that include project-specific risk assessments or method statements (after a site visit from the contractor). These need to be reviewed and agreed by the client representative.

Be aware of contractors who provide you with generic risk assessments. It is not an unknown practice for generic, and often inaccurate, assessments to be passed from contractor to contractor where clients have a requirement for project paperwork, even for large scale projects. Unless the assessments are site-specific, there could be issues that have not been identified at site level that could result in added risk exposure for your organisation.

ILO-OSH 2001	3.10.5.1. / 3.10.5.2 Contracting

See also

Q41	What is risk assessment?
Q73	How do we control our contractors?
Q74	How do we evaluate and select contractors?
Q76	How do we train and instruct contractors prior to commencement?
Q77	How do we manage and supervise contractors?

Q76   How do we train and instruct contractors prior to work commencing?

Organisations (as clients) have a duty of care under occupational safety and health (OSH) legislation to protect third parties, including contractors (this duty of care, of course, also applies to the contractors themselves in the provision of training, information, instruction and supervision to their own staff suitable to their own activities).

Prior to a contractor entering a work location, as the client you already should have a system in place to provide health and safety-related information relevant for the scope of work. Standard practice for many organisations is to use a documented safety induction process that addresses such issues as:

- Layout of the place of work (vehicle parking arrangements, restricted or hazardous areas, etc)
- Operational hazards at the site and those that could impact the contractor.
- Simultaneous operations going on that could impact on the contractor's activities.
- Actions in an emergency and reporting accidents or incidents.
- Contact information for supervisory personnel assigned to the contractor.
- Information on the hours of work, welfare and canteen facilities, etc.

The client also should inform their own employees in good time at locations where contractor work is to take place, informing them of the scope of work and to what extent it may impact their normal day-to-day activities (especially where it could impact on the provision of effective emergency response such as road closures, de-activation of emergency equipment or systems, etc), and put appropriate arrangements in place.

ILO-OSH 2001   3.10.5.1. / 3.10.5.2 Contracting

See also

Q19   How do we define competence?
Q20   How do we develop safety training?
Q57   What is a bridging document?
Q73   How do we control our contractors?
Q74   How do we evaluate and select contractors?
Q75   How do we plan contractor work prior to commencement?
Q77   How do we manage and supervise contractors?

Q77 How do we manage and supervise contractors?

The extent to which a client must manage and supervise a contractor is a combination of many factors, including the type and complexity of the work activity being carried out, the risk to the contractor themselves on the site and the risk to others who may be affected by the contractor's work.

The management and supervision of contractors should take the following into account:

- Any legal requirements or considerations in your jurisdiction that may apply for work activities contracted between a client and a contractor.

- The scope of work and the controls and mitigations must be agreed and understood by both parties prior to work commencing.

- Where deviations from the agreed scope of work are required, which could have a negative health and safety implications, these should be discussed, risk assessed and agreed at management level (this can be handled via the client management of change process where it is available).

- Where contractors are not meeting the agreed requirements for the work, the work should be stopped and discussions held to identify the problems and corrective actions.

Supervision of contractor activities can include:

- Assigned full-time escorts for contractors undertaking short period and / or low risk activities (noise surveys, external management audits, etc).

- Periodic inspections, checks and reviews of contractor work documentation (risk assessments, etc).

- Ensuring complete compliance with the requirements of any permit-to-work system in use.

ILO-OSH 2001   3.10.5.1. / 3.10.5.2 Contracting

See also

Q52   What is a permit-to-work system?
Q73   How do we control our contractors?
Q74   How do we evaluate and select contractors?
Q75   How do we plan contractor work prior to commencement?
Q76   How do we train and instruct contractors prior to work commencing?

EVALUATION

Q78   What are leading indicators?

Leading indicators can be defined as: *"pro-active reports or information that highlight safety issues before they result in accident or incidents"*. A wide range of leading indicators can be defined and measured within a safety management system (SMS). Figures from leading indicators can be used as a measure against your organisation's internal performance criteria or industry-wide safety performance standards, or to meet an industry-specific legal or contractual requirement.

Leading indicators should be:

- Clearly defined within the SMS.
- An objective and reliable metric.
- Easily understood across all departments of an organisation.
- Easy to gather and analyse.
- Communicated and accessible throughout an organisation.

Examples of leading indicators include:

- Number and type of near miss reports.
- Number and category of unsafe act and unsafe condition reports.
- Number of good or safe observations made.
- Number of work procedures reviewed and updated per year.
- Number of risk assessments carried out.
- Number of safety inspections conducted.
- Improvement suggestions from staff (measured in terms of both quality and quantity).
- Number of cross audits (where one department audits another within an organisation).
- Number or percentage of remedial actions closed out on time.

It is very important that the leading indicators you choose actually impact on lagging indicators. Without the cause-and-effect relationship between leading indicator inputs and lagging indicator outputs, the data will have no added value. For example, if there is an issue with eye injuries in the workplace, it makes sense to assess the use of eye protection and then measure that against eye injuries. If eye injuries are not happening, monitoring the use of eye protection will not be of any benefit.

See also

Q3   How is a safety management system organised?
Q78   What are lagging indicators?
Q80   How do leading and lagging indicators work?
Q81   How do we use leading indicators effectively?
Q82   How do we collect leading indicator information?
Q97   What are corrective actions?
Q99   What is continual improvement?

Q79   What are lagging indicators?

Lagging or trailing indicators can be defined as *"re-active information that identifies deficiencies which did result in an accident or incident"*. As such, they provide a measure of how safety has been managed in the past. Figures from lagging indicators can be used as a measure against your organisation's own internal performance criteria or industry-wide safety performance standards[46] or to meet an industry-specific legal or contractual requirement.

Lagging indicators should be:

- Clearly defined within the SMS.
- An objective and reliable metric.
- Easily understood across all departments of an organisation.
- Easy to gather and analyse.
- Communicated and accessible throughout an organisation.

Examples of lagging indicators include:

- Accident reports (lost time injuries, restricted work and medical treatment cases, etc.).
- Property damage reports.
- Environmental leaks or spill reports.
- Accident investigations.
- Accident and incident statistics.
- Dangerous occurrence reports.
- Defective equipment reports.

---

[46] In the oil and gas exploration industry, the International Association of Oil and Gas Producers (OGP) publishes annual industry-wide accident and incident statistics from data submitted by its members. Bear in mind that OGP member organisations' incident reporting is defined to match OGP requirements.

See also

Q80 How do leading and lagging indicators work?

Leading indicators are recognised as one of the foundation stones of preventing reportable and recordable accidents and incidents in the workplace. The principle is simple: that the identification and correction of potential problems at an unsafe act, unsafe condition or near miss level before they can escalate will reduce the number of more serious accidents occurring. Think of the occurrence of a serious accident as a number of escalating events and circumstances happening one after another (similar to a row of dominos falling when the first one is knocked). If you can take corrective action on an unsafe act or condition (remove one or two early dominos), then you break the sequence of events and prevent a more serious accident occurring.

To get the full benefit of these indicators, your safety management system must have a well-defined and robust reporting, tracking and corrective action system in place.

See also

Q3   How is a safety management system organised?
Q78   What are leading indicators?
Q79   What are lagging indicators?
Q81   How do we use leading indicators effectively?
Q82   How do we collect leading indicator information?
Q97   What are corrective actions?

Q81 How do we use leading indicators effectively?

It is important that the use of leading indicators as a measure of safety performance and an opportunity for improvement is targeted appropriately and does not become a numbers game.

Ensure that:

- There is a direct cause-and-effect relationship between leading indicator inputs and lagging indicator outputs.
- The focus is on the quality of reports and the remedial actions taken to correct faults that are identified, not the number of reports.
- The leading indicators are appropriate to the development stage of the safety management system (SMS).

The maturity of a corporate safety culture is important when developing leading indicators, which must recognise the development stage the SMS has reached. There may be variations of the safety culture within a single organisation and at different work locations, so these must be taken into consideration too.

One approach[47] used in the offshore industry is to identify your SMS and, therefore, your safety culture in terms of maturity:

- Level 1 – Compliance.
- Level 2 – Improvement.
- Level 3 – Learning.

Level 1 leading indicators emphasise compliance with the requirements of the SMS and the laws and regulations that apply to operations. The indicators should target the areas where there are the greatest threats to

---

[47] *Leading Performance Indicators – Guidance for Effective Use - Step Change in Safety,* available at http://www.stepchangeinsafety.net/stepchange/News/ StreamContentPart.aspx?.

compliance and which would have the most serious consequences if a failure occurred.

Level 2 leading indicators emphasise identification of areas of weakness and where focused attention is required to bring about improvements. Since historical data on safety performance is available at this level of safety management development, indicators can be used to target areas where performance is below the expected standard.

Level 3 leading indicators emphasise the use of the workforce to develop their own indicators to measure improvements, since continuous improvement and learning now are an established part of the safety culture.

See also

Q78   What are leading indicators?
Q79   What are lagging indicators?
Q80   How do leading and lagging indicators work?
Q82   How do we collect leading indicator information?
Q83   What is occupational safety and health performance?
Q99   What is continual improvement?

Q82 How do we collect leading indicator information?

One of the simplest, but most effective, methods of collecting leading indicator information throughout an organisation is to use what are commonly called 'observation cards'. These are printed cards made freely available at all of your work locations that provide any employee with the opportunity to report on issues and activities that occur during day-to-day operations.

In developing an observation card, take the following into account:

- The range of leading indicators and the classifications used on your cards depends upon the requirements defined within your safety management system.

- Ensure that you have a system to regularly collect, record, collate and analyse the data being made available to you.

- Ensure that, where appropriate, remedial actions are generated for safety critical issues that are raised – this has two immediate benefits: first, employees see that their comments actually drive improvement and, second, they also see management commitment to safety.

- Accept that there will always be a small number of trivial observations made that you will have to filter out.

- Use incentive schemes or programmes to encourage their use such as 'best of the week' or something similar.

- Ensure that your safety training addresses the use and promotion of observation cards.

- Ensure that observation cards provide the opportunity to acknowledge good behaviours and safe practices.

- Ensure that the cards are widely available and that there are plenty of secure boxes or stations to put completed cards into.

- Encourage anonymous reporting of unsafe acts or conditions at your place of work.

- Keep your workforce and employee representatives informed about the results of observation card analysis by providing information in meetings and other safety forums, or on notice boards.

In designing observation cards, consider:

- Size: Keep them conveniently small but not so small that it is difficult to write or read anything on them - A6 or something similar is a good size.

- Material: Print them on cardboard similar to good quality business cards.

- Layout: Use both sides: one side for printing the tick boxes, space for comments, etc. and the other for an explanation of the submission procedure, definition of terms, benefits of reporting, additional notes and so on.

- Content: Include:
  - Observation report card number (for the person collating the cards).
  - Location / Department.
  - Date.
  - Tick boxes for the leading indicators you wish to track (unsafe acts, unsafe conditions, unsafe procedures, good work practices, etc).
  - Description of the observation.
  - Immediate action taken.
  - Remedial actions or recommendations.
  - Name (optional).

The system will take some time to become a normal part of your safety culture but, once in place and properly managed, it can provide invaluable information about what is really happening within your organisation.

A word of warning: Some companies set a number of observation cards that they require their employees to submit per week, per month, or per work rotation, often putting these targets into the corporate or project health and safety objectives. While this ensures that observations are submitted, there may be a issue if employees become obliged to submit cards simply to make up the health and safety statistics.

See also

Q78   What are leading indicators?
Q79   What are lagging indicators?
Q80   How do leading and lagging indicators work?
Q81   How do we use leading indicators effectively?
Q97   What are corrective actions?

Q83   What is occupational safety and health performance?

Occupational safety and health (OSH) performance can be defined as: *"a measure of the level of effectiveness of those business activities aimed at the prevention of injury and disease to persons in the workplace".*[48]

Depending upon what elements of OSH performance are being measured (the primary metrics used generally are based upon the collection of data relating to leading and lagging indicators), the periodic monitoring and review of these should aim to demonstrate:

- Historical OSH performance and performance trends.
- Compliance with statutory obligations.
- Visible commitment to duty of care responsibilities.
- Good management practices.

All organisations should establish procedures for the measurement of OSH performance which should detail responsibility and accountability within the organisation for this process.

See also

Q1   What is health and safety?
Q78   What are leading indicators?
Q79   What are lagging indicators?
Q84   Why should we measure occupational safety and health performance?
Q85   Why should we analyse safety statistics?

---

[48]   Department of Employment and Workplace Relations (Australia) (2005). *Guidance on the Use of Positive Performance Indicators to Improve Workplace Health and Safety*, p.3.

Q84 Why should we measure occupational safety and health performance?

The main reasons for measuring occupational safety and health (OSH) performance are:[49]

- To minimise the occurrence of workplace injury/disease by reducing the level of risk at work: This is perhaps the most important reason to accumulate and analyse OSH performance data within an organisation – that is, to monitor the level of success of existing controls and to highlight areas of improvement for managing workplace risk.

- To provide an OSH feedback mechanism: Once OSH data has been collected, it can be used by both management and workers to measure whether the controls and initiatives implemented to manage workplace risk have been effective.

- To provide a measure of sound management and corporate sustainability: Organisations that require funding from institutional investors are aware that good OSH performance is seen as a positive sign that management has:
  - Sound processes and procedures in place to identify and manage workplace risk.
  - Strong commitment to improving OSH performance.
  - Allocated adequate resources to reducing costs associated with accidents and incidents.

- To facilitate a process of OSH benchmarking between organisations and industries: Organisations in many industry sectors submit data to a central body, usually a membership or trade association, that publishes periodic data on the general level of OSH performance in the industry, thus allowing organisations to measure their own performance against their own industry's

---

[49] Department of Employment and Workplace Relations (Australia) (2005). *Guidance on the Use of Positive Performance Indicators to Improve Workplace Health and Safety*, p.3.

standard. This data tends to be accumulated from lagging indicators, such as the number of fatalities, the number of lost time work days, total recordable incident rates, etc. Leading indicators can be difficult to use for industry sector benchmarking purposes as they tend to be company-specific and tailored to meet individual corporate objectives.

See also

Q34   What are safety management objectives?
Q83   What is occupational safety and health performance?
Q85   Why should we analyse safety statistics?
Q99   What is continual improvement?

Q85   Why should we analyse safety statistics?

Once a safety management system (SMS) has been defined and is run according to the criteria within it (especially the reliable reporting of leading and lagging indicators), the analysis of safety report statistics is a useful tool to measure performance.

Analysis of safety data allows you to address:

- What safety issues are happening in the workplace.
- How well you are doing in managing risk (or not, as the case may be).
- What safety issues have happened in the past.
- Indications that problems are building up and that action needs to be taken.

One of the primary aims of any SMS is that of continual improvement and reporting and analysis of statistical data is one of the main ways to do this. However, some caveats:

- Analysing numbers from your safety activities should only be carried out if your system is well defined and working effectively. For example, it is possible for large organisations to generate very high levels of apparently accident-free exposure man-hours[50] that can provide an illusion of effective safety management but where there is virtually no lagging or leading indicator reporting.
- Whatever criteria are used to analyse accident statistics should be consistent year-on-year to allow for direct and equitable comparison with historical data. That way, improvement in safety performance can be measured.

---

[50] A metric that records the number of exposure hours worked per man, free from reportable accidents, calculated on a 12-hour or 24-hour day. Thus, 25 men working 12-hour shifts for 1 week without an accident is 25 x 12 x 7 = 2,100 man hours.

- When looking at health and safety statistics, always bear in mind that well-known quote: *"There are three kinds of lies: lies, damned lies, and statistics"*.[51]

See also

Q3   How is a safety management system organised?
Q78 What are leading indicators?
Q79 What are lagging indicators?
Q83 What is occupational safety and health performance?
Q99 What is continual improvement?

---

[51] Often attributed to the 19th century British Prime Minister Benjamin Disraeli, although the source is disputed.

Q86   Why do accidents happen?

The primary contributing factors to the cause of accidents in the workplace (or elsewhere) are:

- The behaviours of the human being - it is estimated that over 90% of accidents involve human error.[52]
- The workplace factors that influence human behaviours at work.
- The human factors that influence human behaviours at work.
- The failing of management systems that allow a culture of human error and undesirable workplace and / or human factors to co-exist.

A well-managed workplace has a safety culture – that is, organisational attitudes, beliefs and ways of working that place a high emphasis on safety, such that a careless or irresponsible worker is seen to be behaving outside of the workplace culture and therefore in an unacceptable and unsafe way.

Since workplaces are outside of the control of most workers, management has the responsibility of:

- Organising the safety management system in such a way as to make the careless or irresponsible worker the exception.
- Putting systems in place to modify their unsafe behaviours.

Consider these two scenarios:

1. **Truck-driver A**, driving a new recently-serviced heavy goods vehicle and who has worked for the same company for the last 10 years, has pulled into a motorway service station for lunch, knowing that he has more than sufficient time to make the two more planned deliveries he has (route planning for today meant he took an alternative route to avoid heavy traffic due to road works on the normal route) and be back at the depot by the normal end of the day. He received a bonus payment from

---

[52]   European Agency for Safety and Health at Work (2002). *New Trends in Accident Prevention Due to the Changing World of Work.*

his manager during the recent annual job review for his excellent attitude at work, clean vehicle, tidy personal appearance, for the many positive customer comments about him and for the suggestions for improvement he has made in the past year which have saved his employer money.

2. **Truck-driver B**, driving a 10-year-old vehicle over the speed limit, has to make three more deliveries before he can finish since he was delayed earlier in the day in heavy traffic due to road works on his regular route (why didn't the delivery dispatcher say something about the road-works!). The company usually has too many drops in a day anyway but then always threatens to find other drivers if he says can't do them. The pay is rubbish but it's the only job he could get because his HGV license was suspended due to a drink driving conviction a year ago and anyway, he doesn't pay any tax so he should be grateful. He's not sure about the steering wheel vibrating this much at this high speed and the steering pulling badly to the left under braking, but no one was interested in listening to him about it when he told them last week.

Although contrived, the message should be clear. The second story has all the contributory elements to an accident:

- Human error (violation), by speeding in his truck.
- Workplace factors (such as poor vehicle maintenance and unsafe work equipment).
- Personal factors (such as bullying and abuse).
- Human factors (fatigue and stress over the threat of losing his job, sitting in a traffic jam and having to meet unrealistic delivery schedules).

Can you see the direction an accident investigation may take if an accident happened?

ILO-OSH 2001	3.12. Investigation of work-related injuries, ill-health, diseases and incidents, and their impact on safety and health performance

See also

Q87   What human errors are contributory factors to accidents?

Accidents almost always are caused by human involvement or intervention, very often by the carelessness, stupidity or incompetence of the people involved. However, a systems approach looks beyond the obvious causes to identify where human error has contributed to an accident, perhaps because the systems were not in place to prevent them or because they failed in some way.

Human error can contribute to accidents in four ways:[53]

Human error category	Definition
Slip	When a person does something but it is not what they meant to do.
Lapse	When a person forgets to do something due to a failure of attention / concentration or memory
Mistake	When a person does what they meant to do, but should have done something else.
Violation	When a person decided to act without complying with a known rule, procedure or good practice.

The first three categories normally are the outcome of a poor safety management culture, the result of 'slippage' in the effective management of the workplace influenced by often undesirable workplace factors combined with low level human failures.

The 'violation' category is different, since it implies a deliberate act of non-compliance and, therefore, often is seen as the responsibility of the individual (who must be held accountable for their actions). Although

---

[53]   Energy Institute (2008). *Guidance on Investigating and Analysing Human and Organisational Factors Aspects of Incidents and Accidents*, London.

recognising that the degree of violation may vary (5kph over the speed limit is a lesser violation than 60kph over, for example), an organisation must ensure that its systems identify and correct violation behaviours to prevent a situation where minor violations become accepted as normal.

See also

Q86 Why do accidents happen?
Q88 What workplace factors are contributory factors to accidents?

Q88   What workplace factors are contributory factors to accidents?

Workplace factors are those identified as having a significant impact on human behaviours, including environmental factors:

- General workplace factors:
    - No safety, hazard or other workplace signage.
    - Lack of proper maintenance of work equipment.
    - Use of unsafe work equipment.
    - Poor communications on the shop-floor.
    - Poorly-designed work areas.
    - Too noisy, too cold or too hot working environment.

- Personnel factors:
    - Lack of training.
    - Incompetent management.
    - Bullying and abuse.
    - Low morale due to low pay and poor working conditions.
    - No career development.
    - Use of unqualified personnel.

- Organisational factors:
    - Lack of adequate supervision.
    - Lack of a developed safety management system.
    - Management not committed to effective safety management in the workplace.
    - Management not complying with the most basic health and safety legal standards.
    - Lack of, or non-use of, a disciplinary procedures.
    - Obvious violations from management seen by the workforce.

- Task factors:
    - Repetitive and boring work.
    - Lack of job rotation.
    - Use of old and unsafe work equipment.
    - Requirement to work long hours.

- No work procedures or poorly developed / inadequate procedures.

Other factors include:

- Operational change: Does your organisation have the flexibility and the safety management structures in place to identify, assess and manage planned and unplanned changes to your operations? Does your organisation ignore new risks, continuing to concentrate on already well-established operational risks?

- Organisational change: Downsizing, take-overs, joint ventures and other major organisational changes that occur totally outside of the control of the worker can have an effect. Studies have identified links between downsizing/organisational restructuring and increased occupational violence, bullying or aggressive behaviour at work.[54]

- Contract and temporary workers: Due to the often precarious nature of their employment, contract and temporary workers tend to have higher accident rates than full-time employees. Do you analyse who are the people involved in accidents and could this be a contributory factor in your workplace?

- Increased mobility: Both workers and goods travel much more now. Is your organisation susceptible to the increased risk of road traffic accidents caused by the illegal use of mobile phones in vehicles, lack of emergency planning in the event of an accident, lack of journey management or route planning, plans for travel (or not) in severe weather conditions and other factors that can increase risk to people and goods?

- Size: Accident risks are higher for those employed in small and medium enterprises (SMEs) where the incidence rate for fatal accidents to workers in enterprises with less than 50 employees is

---

[54] European Agency for Safety and Health at Work (2002). *New Trends in Accident Prevention Due to the Changing World of Work.*

about twice the rate of larger units.[55] Are you in this more vulnerable group?

- Staffing arrangements and manning: Could a change (reduction, re-organization or redistribution) of employees in your workforce place new stresses on the remaining workforce in meeting targets, having to working with new untrained people or working with fewer personnel to meet the same workload?

- Fatigue and stress: Are you aware of requirements for workers to work excessive overtime (low base salary but high overtime rates) or work irregular hours due to shift work, especially night shifts? Are workers under stress from financial pressure due to family circumstances or other non-occupational issues in their lives? Is bullying in your workplace causing stress?

ILO-OSH 2001	3.12. Investigation of work-related injuries, ill-health, diseases and incidents, and their impact on safety and health performance

See also

Q59   What is management of change?
Q86   Why do accidents happen?
Q87   What human errors are contributory factors to accidents?

---

[55]   European Agency for Safety and Health at Work (2002). *New Trends in Accident Prevention Due to the Changing World of Work.*

Q89   Why do we investigate accidents?

This is a simple question with a complicated answer. We should investigate accidents in the workplace so that: *"an organisation can fulfil its legal and moral obligation to identify failures in the workplace under a duty of care, to implement controls to eliminate or more effectively manage them, to reduce personal suffering and to more clearly understand the economic benefits of preventing further accidents"*.

Accidents are a learning experience, albeit with a high price. If you can identify the reasons why an accident occurred, preventative actions put into place to correct failings should reduce the risk of the same or similar loss happening again.

Reasons for investigating accidents include:

- Duty of care.
- Moral imperative.
- Financial cost.

Most jurisdictions place a duty of care on the employer to provide a safe place of work, safe work equipment, personal protection equipment, etc. Investigating accidents, especially after an injury or fatality, is part of that duty of care in order to prevent a re-occurrence. Increasingly, it is becoming a legal requirement in many jurisdictions (ensure you are aware of the legal requirements to investigate accidents in your operating location(s)).

Families can suffer serious long-term disruption to their lives where the main 'breadwinner' is permanently injured or killed at work. Unfortunately, this is still a reality for many thousands of families every year across the industrialised world – for example, across Europe, over 5,500[56] people are killed in workplace accidents every year. The physical and mental toll on families that can result from a serious workplace

---

[56]   European Agency for Health and Safety at Work (2002). *Inventory of Socioeconomic Costs of Workplace Accidents*.

accident to a family member cannot always be measured simply in financial terms.

Accidents cost businesses money. In the event of an accident (depending on the scenario), costs include:

- Loss of production.
- Loss of reputation (and therefore potential loss of future orders).
- Re-training of staff.
- Cost of hiring temporary workers.
- Buying new equipment to replace damaged units.
- Increased insurance premiums.
- Costs related to court cases and litigation under health and safety legislation, including fines.
- Costs related to court cases and litigation where business owners may be sued for compensation in the civil courts.

In the EU, estimated costs to Member States[57] due to accidents in the workplace vary from 1% to 3 % of gross national product.

ILO-OSH 2001	3.12. Investigation of work-related injuries, ill-health, diseases and incidents, and their impact on safety and health performance

See also

Q90   How do we investigate accidents?
Q91   How do we prevent accidents?
Q97   What are corrective actions?

---

[57]   European Agency for Health and Safety at Work (2002). *Inventory of Socioeconomic Costs of Workplace Accidents.*

Q90   How do we investigate accidents?

Conducting accident investigations in the workplace can be difficult and trying, especially in circumstances where more serious outcomes have occurred.

Whatever specific methodology you choose, you should use a 'system-based' approach, the objective of which is not only to establish the 'what' and the 'how' of an accident but, most importantly, the 'why'. Unless you identify and address the true underlying causes of an accident during an investigation, the accident may be repeated under different circumstances but still caused by underlying uncorrected system failures.

Management commitment to the investigation process is vital and it is an organisation's management responsibility to:

- Be committed to, and support, the investigation process.
- Provide the required resources for the accident investigation and the follow-up of remedial actions.
- Be willing to learn from failings in the system and mistakes that have been made.
- Be open and honest on findings where failings in the safety management system (SMS) have been identified.

The competency of accident investigation team members is also important. You should define the minimum training requirements for your own employees to act as accident investigation team members and, where these are not / cannot be met, to bring in experienced health and safety professional(s) as external competent person(s) to lead an investigation.

The standard workflow for an accident investigation process is based on five elements:

- Reporting the accident or incident.
- Conducting an investigation.
- Generating recommendations (actions).

- Tracking remedial actions.
- Sharing information.

Reporting of accidents and incidents must be defined in your SMS. There should be clear guidance as to the requirement for an accident investigation, how to classify accidents (as to their severity and seriousness), what the legal reporting requirement is in your local jurisdiction, and who will be on the investigation team. Consideration needs to be given to joint investigations (in the event of a serious accident) with regulatory health and safety authorities and even the police, depending on the circumstances. Does your safety representative have a defined legal role to play in an investigation and, if so, do you understand what that role is?

The level of investigation, the composition of the investigation team and the resources to be made available depends on the severity of the accident. The general workflow for an investigation should be to establish:

- What happened.
- Who was injured, what was damaged and to what extent.
- What were the environmental conditions at the time.
- What were the time-lines for the accident.
- What was the chain of events prior to the accident.
- What activities were going on at the time.
- What documents, procedures or records are available, if any.
- Whether there were any unusual or abnormal conditions, work activities or circumstances present at the time.

Last, the investigation should include witness interviews, which must be handled sensitively, especially in the event of a fatality or the requirement to interview injured persons. Specialist guidance is recommended for this part of the process if your organisation does not have already clearly defined guidelines and experienced investigators available.

Depending on the outcome of the investigation, recommendations (or actions) must be developed for improving the system and for correcting deficiencies. The actions developed should use the SMART criteria: **S** – Specific; **M** – Manageable; **A** – Achievable; **R** – Realistic; and **T** – Time-bound. Actions must be assigned to responsible people with the appropriate level of authority and access to resources to get things done.

Learnings from accidents can provide useful information internally and to other organisations, especially those in similar industries. This can be done internally via notice boards and reviews at safety meetings, and externally via safety alerts and bulletins issued through industry associations.

ILO-OSH 2001	3.12. Investigation of work-related injuries, ill-health, diseases and incidents, and their impact on safety and health performance

See also

Q3   How is a safety management system organised?
Q19   How do we define competence?
Q89   Why do we investigate accidents?
Q91   How do we prevent accidents?
Q97   What are corrective actions?
Q99   What is continual improvement?
HSE Books (2004). *Investigating Accidents and Incidents: A Workbook for Employers, Unions, Safety Representatives and Safety Professionals* (HSG245).
Energy Institute (2008). *Guidance on Investigating and Analysing Human and Organisational Factors Aspects of Incidents and Accidents.*

Q91   How do we prevent accidents?

Accident prevention is a natural outcome for an organisation that improves the way risk is managed in the workplace, normally within the framework of a safety management system.

"The strong economic advantages of good occupational health practice need to be highlighted continuously to organisations because the failure to acknowledge the importance of this link will limit the effectiveness of interventions aimed at preventing disease and injury".[58]

The accident prevention strategy that an organisation operates depends on:

- Contributory factors (human error, human and workplace factors) present due to the scope, size and complexity of your operations.

- The effectiveness of the procedures that you have in place for:
  - The reporting and tracking of accidents and incidents.
  - The analysis of trends for all leading and lagging indicators that can be used to identify the vulnerabilities and weaknesses within your organisation.
  - The management of remedial actions.

- The resources that the organisation is willing to make available for occupational health and safety.

- Last, but certainly not least, the commitment from senior management to the ongoing reduction of accidents and incidents in the workplace, year on year.

ILO-OSH 2001	3.12. Investigation of work-related injuries, ill-health, diseases and incidents, and their impact on safety and health performance

---

[58]   Lahiri, Levenstein, Nelson and Rosenberg, 2005; Toffel and Birkner, 2002.

See also

Q92   What is an audit?

An audit of an occupational safety and health (OSH) management system can be defined as: *"an evaluation of an organisation's OSH management system elements or a subset of these, in order to determine whether the system and its elements are in place, adequate, and effective in protecting the safety and health of workers and preventing incidents".*[59]

Audits are a vital component in improving your safety management system (SMS) over time because they allow your organisation to identify and assess:

- Where improvements are required in the system.
- Where there are omissions in the system.
- Whether the system is fit for purpose in meeting company policies and OSH objectives.
- Whether the system is meeting regulatory and management system standards.
- Whether the allocation of resources is adequate and well-targeted.

The basic audit framework and criteria should be outlined clearly within your SMS, and should include:

- Definition of auditor competency (both for internal and external audits).
- Definition of persons responsible in line management for ensuring that audits take place.
- The scope of individual audits (especially where there may be a range of audits defined within your SMS).
- The frequency of audits.
- Audit methodology and reporting requirements.

---

[59]   *ILO-OSH 2001* – Section 3.13, Audit.

Careful consideration must be given to the structure, requirements and standards of your SMS when developing an effective audit protocol for your organisation, including:

- What do we audit against?
- How thorough do we need to be?
- How often do we need to audit?
- What is the audit process going to be?
- What outputs and records are generated for audits?

The inspection process usually is based upon a checklist or question approach and is designed to be quite prescriptive, to get a specific answer to a specific question. In contrast, the audit process is a much more involved process which requires:

- Planning.
- Allocation of resources.
- Conducting the audit:
  - Interviewing and talking to staff.
  - Watching workplace activities and practices.
  - Looking at the relevant documentation (depending on the audit scope)
- Close-out.

ILO-OSH 2001	3.13. Audit

See also

Q3	How is a safety management system organised?
Q93	What do we audit against?
Q94	How thorough do we need to be in an audit?
Q95	How often do we need to audit?
Q96	What is a management review?
Q99	What is continual improvement?

Q93   What do we audit against?

The safety management system (SMS) that your organisation implements should reflect the regulatory environment, industry standards that apply to your organisation as well as its own policy objectives. Therefore, the starting point for an audit is to identify what these are:

- Regulatory standards: What laws and regulations apply to your organisation in the jurisdiction(s) that you operate in? Identify the basic occupational health and safety Acts and regulations which cover all employers. Next, are there specific regulations that apply to your industry sector, such as food hygiene (HACCP),[60] the management of major accident hazards (COMAH),[61] etc? Are there Codes of Practice (CoP) or Approved Codes of Practice (ACoP) developed for specific work activities that apply to your activities? If so, ensure your audit is tailored to identify these compliance requirements.

- Industry standards: If your organisation subscribes to an industry standard such as OGP[62] or IAGC[63] (as used in the offshore and geophysical exploration industry sectors), you must take their guidelines into consideration when developing audit content. It may be possible to use the industry standard formal audit

---

[60]   Hazard Analysis and Critical Control Points – This regulation derives from EU Council *Directive 93/43/EEC* of 14 June 1993 on the hygiene of foodstuffs. In Ireland, for example, this directive is set in law as *European Communities (Hygiene of Foodstuffs) Regulations, 2000*.

[61]   The *Control of Major Accident Hazards Regulations, 1999* as amended by the *Control of Major Accident Hazards (Amendment) Regulations 2005*, implement the EU *Seveso II Directive 96/82/EC* as amended by *Directive 2003/105/EC* in Great Britain

[62]   International Association of Oil and Gas Producers – "The International Association of Oil & Gas Producers has access to a wealth of technical knowledge and experience with its members operating around the world in many different terrains. We collate and distil this valuable knowledge for the industry to use as guidelines for good practice by individual members".

[63]   International Association of Geophysical Contractors.

documentation directly, without developing your own content against these standards.

- Voluntary standards: If your organisation has adopted any ISO[64] or OHSAS[65] quality or management standard, these must be incorporated into your audit framework, although in these cases, an external organisations (such as a certification body) may be responsible for carrying out the audit.

- Policy and objective standards: What detail within your own SMS needs to be examined in an audit? For example, in a company with a stated substance abuse policy with a drug and alcohol testing regime, the audit should identify this requirement, then examine the issues that can impact the effectiveness of this policy, including:
  - Have drug and alcohol tests been carried out according to documented procedures?
  - Is there a minimum competency and training requirement for persons undertaking drug and alcohol tests, is it defined and do testers meet that requirement?
  - Is there an effective 'chain of custody' defined for test samples?
  - Does the procedure meet the current legal requirement for testing of substances at work?
  - When was this policy statement and procedure last reviewed?

For example, a company has decided to use full body harnesses for working-at-height, which are manufactured to standard BS EN 361:2002. The company decides to adopt the manufacturer's statement of obsolescence (which outlines the expected lifespan of the product) as their own company standard for keeping such equipment. So the company requirement is detailed in a management procedure as:

- Once the product is taken from this original packaging for the first time, this date becomes the 'date of first use', which should be recorded on the Company Inspection Register and the 4-year working life begins.

---

[64] International Standards Organisation.
[65] Occupational Health and Safety Administration System.

- Any item of fall protection equipment is subject to a maximum working life of four years from the recorded date of first use (provided that the item has been correctly stored, maintained and subjected to regular inspections by a trained and competent person).

- A new item of fall protection equipment may be stored for a maximum of three years and will still give the potential four-year working life, provided it remains in the original manufacturer's packaging.

If this is the company standard, the audit process must check for conformance to this management system requirement. So the audit of working-at-height gear, which is undertaken on a quarterly basis, should include:

- Verify that a Company Inspection Register is available and is in use.

- Are all items of fall protection equipment in use during the audit recorded in the Company Inspection Register?

- Are any items of fall protection equipment in use during the audit beyond their four-year 'date of first use' limit?

- Are any stored items of fall protection equipment beyond their three-year storage limit?

- Are all stored items in their original packaging?

So, five simple questions confirm that this part of the SMS meets the company-specified requirements. It is this simple principle (check whether you do what you say you do) on which an organisation can start to compile an internal audit programme to reflect its particular reality.

ILO-OSH 2001   3.13. Audit

See also

Q3   How is a safety management system organised?
Q7   What is a policy statement?

Q94   How thorough do we need to be in an audit?

Rather than thinking of thoroughness in terms of an individual audit (is it detailed enough?), think of ensuring that the scope of audits and inspections that can be implemented within your organisation over a year will provide an accurate measure and thorough assessment of the state of your safety management system.

Generally speaking, the more frequent audits and inspections are (daily, weekly, monthly), the more focused and quicker they should be and the less frequent they are (quarterly, biannual and annual), the more wide-ranging (and, therefore, time-consuming) they should be, so aim for a balanced combination of the two.

In addition to an internally-developed audit programme, you must incorporate any regulatory or industry-specific audits into the programme where there is a requirement on you to do so.

It is also important to appreciate that there are limitations in the audit process too, often due to time pressure and a lack of resources, so it is necessary to be realistic in what you expect the audit process to achieve within your organisation (or how thorough you can afford to be):

- **Time pressure:** In business, time costs money and whether you have external auditors on site or use your own personnel for the audit process, they have a direct cost to your business. The quicker an audit can be carried out, generally the cheaper it will be, so there is always a balance to be achieved between what you expect from an audit and how much time you are willing to pay for.

- **Lack of resources:** Sometimes, for more significant audits (such as management reviews, etc), organisations can be unrealistic about what an auditor can achieve within a given time frame. For example, the outcome of a management review will be different if two people are working on it for a week, rather than one person for three days.

In the end, you get out of the audit process what you are willing to put into it in terms of time and resources.

ILO-OSH 2001	3.13. Audit

See also

Q92 What is an audit?
Q93 What do we audit against?
Q95 How often do we need to audit?

Q95   How often do we need to audit?

A better question is 'what audits and inspections will occur and when?'.

Your starting position is to look at the audits and inspections required under the law so that you can incorporate these into your own audit programme. There will be a great deal of variation here, depending on:

- What jurisdiction you are in and, therefore, what general health and safety legislation applies.
- Whether there are any specialist regulations that apply.
- What activities you undertake and what equipment you are working with.
- Whether you subscribe to any dedicated programmes, systems (such as HACCP) or standards as part of your safety management system that have compulsory audit or inspection elements within them.

ILO-OSH 2001	3.13. Audit

See also

Q3	How is a safety management system organised?
Q92	What is an audit?
Q93	What do we audit against?
Q94	How thorough do we need to be in an audit?

Q96   What is a management review?

A management review is a process designed to assess the effectiveness of an organisation's safety management system, and should be carried out by senior management. It should be defined in an organisation's safety management system (SMS) with an detailed process, allocated responsibilities, schedules and the resources available to ensure the review process is effective.

The primary aims of the review process are:

- **Evaluation:**[66]
    - The overall strategy of the SMS, to determine whether it meets planned performance objectives.
    - The SMS's ability to meet the overall needs of the organisation and its stakeholders, including its workers and the regulatory authorities.
    - The need for changes to the SMS, including policy and objectives, to ensure progress towards the organisation's OSH objectives and corrective action activities.
    - The effectiveness of follow-up and remedial actions from earlier management reviews.
- **Identification:**
    - What action is necessary to remedy any deficiencies in a timely and effective manner.
- **Provision:**
    - The feedback of findings into the system and the assessment of priorities, for meaningful planning and continual improvement.

In addition, consider:

- **Timing:**
    - The frequency and scope of periodic management reviews should be defined according to the organisation's needs and conditions.

---

[66]   *ILO-OSH 2001* – Section 3.14, Management Review.

- **Inputs to the system:**
  - The analysis of leading and lagging indicators such as work-related injuries, ill-health, diseases and incident investigations.
  - Performance monitoring and measurement, audit activities and additional internal and external inputs.
- **Communication:**
  - The findings of the management review should be recorded and formally communicated to the safety and health committee, workers and their representatives and to those who are responsible for the relevant element(s) of the OSH management system so that they may take appropriate action.

The management review process is an oversight of the SMS, whereas the audit process tends to be focused on examining particular elements within the system and is a contributing factor to the management review.

ILO-OSH 2001	3.14. Management Review

See also

Q3   How is a safety management system organised?
Q83   What is occupational safety and health performance?
Q92   What is an audit?
Q97   What are corrective actions?

ACTION FOR IMPROVEMENT

Q97   What are corrective actions?

Preventive and corrective actions[67] are generated within an effective safety management system (SMS) as a significant part of the continual improvement process and can be defined as: *"agreed actions taken to eliminate an occurrence (preventative) or recurrence (corrective) of an unsafe act, unsafe condition or other undesirable situation"*. These actions can be generated from safety management activities, such as:

- Internal audit reports (generated from audit activities from within the organisation, such as cross audits or department audits).
- External audit reports (generated from an external third party assessment of the organisation).
- Accident and incident investigations.
- Near-miss reports.
- Unsafe act and unsafe condition observation reports.
- Safety meetings.
- Management reviews.

Of course, the effectiveness of preventive and corrective actions in the continual improvement process depends on:

- The effectiveness of the systems in place to ensure that responsibilities and accountabilities are clearly defined to make sure that corrective actions are taken care of (do job descriptions within the organisation specifically state that incumbents are responsible for managing corrective actions assigned to them?).
- Adequate resources available to ensure that corrective actions are completed.
- A system is in place to monitor the effectiveness of managing corrective actions.

---

[67] Corrective actions also are known as remedial actions, action items and non-conformity reports although the term 'non-conformity' can have specific legal definitions in some regulations.

Corrective actions are a useful metric for organisations to measure corporate and individual performance. For example, in developing annual safety objectives (either for their organisation as a whole or for a specific project), you could develop a safety objective that: *"75% of all corrective actions should be closed out by the original assigned close-out date"*. In the annual management review, this objective would be assessed and, if not met, recommendations made or, if met, safety incentives or rewards given to the responsible person(s). The success rate for the management of corrective actions could be used as part of a safety performance review for individual managers.

ILO-OSH 2001	3.13.6. Audit 3.14.1d. Management Review 3.15. Preventive and Corrective Actions

See also

Q3   How is a safety management system organised?
Q34   What are safety management objectives?
Q84   Why should we measure occupational safety and health performance?
Q92   What is an audit?
Q96   What is a management review?
Q98   How do we manage corrective actions effectively?
Q99   What is continual improvement?

Q98   How do we manage corrective actions effectively?

Once preventive and corrective actions have been generated (from the many activities of the safety management system (SMS)), the next step is to ensure that these actions are dealt with in a timely and efficient manner. This is normally achieved by creating a central register for all actions generated, often called a 'Corrective Action Plan' (CAP).

Management of the CAP depends on how the SMS has been developed, whether it is part of a customised software application, a spreadsheet application or a paper system. Regardless of how the CAP has been implemented, there should be a management procedure to explain how it is organised and managed.

Corrective action plans often are customised to the specific requirements of the organisation, but the following elements should be considered as the minimum:

- Item number: All actions should be assigned a unique item or action number, even across large organisations and business units where a common register is used.

- Action date: The date that the action was generated. This date is easy to identify from near-miss reports, accident investigations and unsafe act / condition reports, but may be more problematic from audits that run over several days and where it may not be possible to identify a specific day for a specific action. In this case, use the date of the audit closing meeting.

- Observation: Comments on the original observation made on the report card or audit that requires corrective action.

- Action: The specific corrective action required to eliminate or manage the issue noted in the observation.

- Department: For organisations with several internal departments, or for CAP plans that are set up for projects and where a number of contractors may be involved, this section can be used to define which department / contractor is responsible for an action.

- Assignee or responsible person: The name of the person who has been assigned the responsibility to manage the corrective action. It is important to note that this person should have the authority and access to available resources to be able to deal with the corrective action effectively.

- Target date: Target date is the date when the action should be closed out. It can be difficult to assign specific dates for actions, especially where there could be an unknown lead time in ordering equipment or some research is needed to find the best solution to a problem. In such cases, a default time period can be assigned to actions initially – for example, four weeks from the action date – and adjust as necessary.

- Source (Forum): Since a comprehensive SMS generates actions from a wide range of sources, it is useful to track these. It is not always obvious from the observation text where the action came from (such as an unsafe condition report) so detailing the source is vital, especially from audits where it may be necessary to review the audit report itself to understand the detail or context of an action. Examples of sources relating to the forum include observation cards, safety meetings, audit and inspections, etc.

- Source (Location): For organisations with multiple work sites or locations, it is useful to identify which physical locations are generating input into your system. Examples of sources relating to location include factory, warehouse or office locations, business units, operational teams, etc.

- Priority: It is inevitable that corrective actions will need to be prioritised by assigning a priority tag such as 'Urgent, High, Medium or Low', or some other prioritising system. Whatever system is used, define what these terms mean – for example, 'High' may be defined as 'Must be closed out within 7 working days' and 'Medium' as 'Must be closed out within 1 month' or whatever is suitable and appropriate.

- Date closed: The date that the action has been dealt with and is no longer active.

- Status: The status of a corrective action can change over time, so an organisation should decide on what tags are to be used to identify the status of corrective actions. Examples include:
  - Open / Work-in-Progress (WIP) / Active – a corrective action that is still being dealt with.
  - Overdue – a corrective action that is still active or open beyond the stated target date. The assignee must not allow an action to become overdue, but should assign a new target date if there is a valid reason why the original target date is no longer achievable.
  - Rejected – a corrective action that has been rejected. Actions can be rejected by the responsible person if there is a legitimate reason to do so.
  - Closed – a corrective action that has been completed to the satisfaction of the responsible person and verified.
  - Planned – a corrective action that is still active but cannot be completed until some other significant scheduled activity in the future (such as a maintenance shut-down or refit, etc.).
- Comments: Comments should be added by the responsible person on the progress of the corrective action. For example, if a spare part has been ordered to fix a corrective action, a comment outlining the order / requisition number or purchase order placed helps to keep everyone informed as to the ongoing status of the action.

Whatever corrective actions are decided on, ensure that they are SMART,: **S** – Specific; **M** – Measurable; **A** – Attainable; **R** – Realistic; and **T** – Time-bound.

| *ILO-OSH 2001* | 3.13.3. Audit |
| | 3.15.1b. Preventive and Corrective Actions |

See also

Q3   How is a safety management system organised?
Q97   What are corrective actions?

Q99   What is continual improvement?

Continual improvement is the process of taking all the data generated from your operating safety management system (SMS), reviewing it against the internal and external standards that you use and then making changes to your system to improve it. Once your SMS has been defined and is fully operational for a period of time, you should be generating a wide variety of data in a number of ways, all of which can contribute to the improvement process.

The continual improvement process should be defined within your own organisation and should outline which elements of your safety management system are included. You may expect to see some or all of the following (or generate your own additional items) as possible components of this process:

- Review of policy statement objectives and aims.
- Review of occupational safety and health objectives.
- Analysis of leading and lagging performance indicators.
- Analysis and review of risk assessments.
- Review of remedial actions from all audits, inspections and checks.
- Conclusions and findings from accident and incident investigations.
- Feedback from employees (from forums such as safety meetings, safety committees, suggestions for improvement and safety representatives, etc).
- Conclusions from management reviews.
- Reviews of procedures and documentation for compliance with new and evolving Acts, regulations, industry Codes of Practice and industry standards.
- Reviews of procedures and documentation to ensure continuing compliance with, and relevance to, your operations.
- Conclusions and findings on the effectiveness of occupational health promotion schemes.

- Review of training standards (or delivery methods) applicable to your industry sector.

- Periodic performance reviews of employees where health and safety compliance is included.

ILO-OSH 2001	3.16. Continual Improvement

See also

Q3	How is a safety management system organised?
Q34	What are safety management objectives?
Q41	What is risk assessment?
Q78	What are leading indicators?
Q79	What are lagging indicators?
Q90	How do we investigate accidents?
Q96	What is a management review?
Q97	What are corrective actions?
Q100	Why is continual improvement important?

Q100   Why is continual improvement important?

From the first Model T to new electric or hydrogen-powered cars, from the Wright Brothers' first flight to Concorde and from the first Sputnik satellite to the International Space Station, striving for continual improvement is a major component of the human condition. In the 21st century, we do not have to look too far to see the continual improvement process in action. We change our consumer goods on a regular basis as the relevant technologies continually improve, and we have an assumption that the services that are provided to us (such as hospitals, public transport, the road network, schools, etc.) also will evolve, delivering improved performance and higher standards year on year. Indeed, we are so used to it being a part of our everyday lives, we may not even realise how embedded the improvement process is into all aspects of our society.

Your organisation's safety management system (SMS) is the same – without an ongoing continual improvement process, your system will stagnate, become outdated and, in all likelihood, will start to deviate from legal compliance requirements, which negates the whole purpose of having a SMS in the first place.

So, as the world continues to change around us all, your SMS also must continually improve to keep up with those changes. For each and every SMS, there is a need to identify and document which internal and external elements are critical in the improvement cycle, to learn from them and then to implement change.

ILO-OSH 2001	3.16. Continual Improvement

See also

Q99   What is continual improvement?

ABOUT THE AUTHOR

ANDY TILLEARD is a Chartered Member of the UK Institute of Occupational Safety and Health (IOSH), a Full Member of the International Institute of Risk and Safety Management (IIRSM) and is registered as a European Safety and Health Manager as outlined by the European Network of Safety and Health Professional Organisations (ENSHPO).

ABOUT THE QUICK WIN SERIES

The **Quick Win** series of books, apps and websites is designed for the modern, busy reader, who wants to learn enough to complete the immediate task at hand, but needs to see the information in context.

Topics published to date include:

- QUICK WIN MARKETING.
- QUICK WIN DIGITAL MARKETING.

Topics for publication in 2010 include:

- QUICK WIN ECONOMICS.
- QUICK WIN MEDIA LAW IRELAND.
- QUICK WIN LEADERSHIP.
- QUICK WIN LEAN BUSINESS.
- QUICK WIN SALES.
- QUICK WIN SMALL BUSINESS.

For more information, see **www.oaktreepress.com**.

Lightning Source UK Ltd.
Milton Keynes UK
30 September 2010

160611UK00001B/6/P